Cropping Strategies for Efficient Use of Water and Nitrogen

Cover Design: Patricia Jeffson

American Society of Agronomy, Inc.
Crop Science Society of America, Inc.
Soil Science Society of America, Inc.
677 South Segoe Road, Madison, WI 53711 USA

Library of Congress Cataloging-in-Publication Data

Cropping strategies for efficient use of water and nitrogen: proceedings of a symposium sponsored by Divisions S-4, S-6, S-8, C-3, and A-6 of the Soil Science Society of America, Crop Science Society of America, and the American Society of Agronomy in Atlanta, GA, 30 Nov. and 1 Dec. 1987 / editor, W.L. Hargrove.
p. cm. — (ASA special publication; no. 51)
Includes bibliographies
ISBN 0-89118-097-4
1. Cropping systems. 2. Crops and water. 3. Crops and nitrogen. I. Hargrove, W.L. (William Leonard). II. Soil Science Society of America. III. Crop Science Society of America. IV. American Society of Agronomy. V. Series.
S1.A453 no. 51
[S602.5]
631 s—dc19 88-21878
[631.5'8] CIP

Printed in the United States of America

Cropping Strategies for Efficient Use of Water and Nitrogen

Proceedings of a symposium sponsored by Divisions S-4, S-6, S-8, C-3, and A-6 of the Soil Science Society of America, Crop Science Society of America, and the American Society of Agronomy in Atlanta, GA, 30 Nov. and 1 Dec. 1987.

Editor
W. L. Hargrove

Editorial Committee
W. L. Hargrove, co-chair
B.G. Ellis, co-chair
T. Cavalieri
R. C. Johnson
R. J. Reginato

Organizing Committee
W. L. Hargrove
A. L. Black
J. V. Mannering

Editor-in-Chief ASA
G. H. Heichel

Editor-in-Chief CSSA
C. W. Stuber

Editor-in-Chief SSSA
D. E. Kissel

Managing Editor
Sherri Mickelson

ASA Special Publication Number 51

American Society of Agronomy, Inc.
Crop Science Society of America, Inc.
Soil Science Society of America, Inc.
Madison,Wisconsin, USA

1988

CONTENTS

FOREWORD

Water and N are two factors that have a major influence on crop production. Water is provided by rainfall, and where rainfall is insufficient, supplemental irrigation is used. Similarly, N needs can be provided by using legumes and other N_2-fixing systems, or supplemented with N fertilizer and manures. Since crop production is at the beginning of most human food chains and is a essential human activity, water and N influence human health, welfare, and economics in critical ways. Therefore, it is important to understand the interaction of water and N in cropping systems, responses to these vital inputs, and ways of managing water and N most efficiently.

Cropping Strategies for Efficient Use of Water and Nitrogen deals with water and N management from several perspectives, including crop species, crop sequences, cultural practices, inputs, and environmental quality.

We greatly appreciate the contributions of W. L. Hargrove and B. G. Ellis, co-editors, and their editorial committee. We are also grateful to W. L. Hargrove, A. L. Black, and J. V. Mannering for organizing the symposium in which the papers that later were developed into chapters for this book were first presented. We wish to thank the authors for sharing their knowledge, experience, and insights on this subject.

This book is a valuable contribution to scientific, technological, and practical understanding of an important topic. We think you will find it informative and stimulating.

D. A. Holt, *president*
American Society of Agronomy

C. J. Nelson, *president*
Crop Science Society of America

D. R. Keeney, *president*
Soil Science Society of America

PREFACE

The very existence of the human race on this earth depends upon good management and wise stewardship of natural resources to produce food and fiber. The United States has been at the forefront of the science of increasing food and fiber production and as a result has achieved an abundance of food that is cheap, and a high standard of living. But, environmental problems, increasing public concern about degradation of groundwaters, and an economic dependency upon fossil fuels have also resulted. Concern for efficient use of N and water, while still protecting our environment, led to the development of a symposium at the 1987 Annual Meetings of the American Society of Agronomy at Atlanta, GA. The symposium was co-sponsored by Div. S-4, S-6, S-8, C-3, and A-6. Twelve papers were presented which address many factors that affect the efficiency of water and N use.

Cropping Stretegies for Efficient Use of Water and Nitrogen is the written publication from the papers that were presented at the symposium. Papers discuss the use of cropping systems to gain N_2 fixation and efficiency of utilization of this fixed N. Cropping systems, including rotations, multiple cropping, intercropping and specific cropping systems for low input agriculture as well as using specific cropping systems for reduced leaching of salts are topics included in the papers. The keen interest in groundwater quality and in reducing crop production costs while maintaining acceptable yield make the publication a timely contribution.

We, as members of the society, are indebted to the authors whose work has collected thoughts, concepts, and data that relate to this important topic. It is our hope that this will stimulate thinking about research that is essential for more efficient use of two of our most important resources, N and water, in our agricultural production systems. We wish to thank the reviewers who contributed suggestions for improvement of the manuscripts and for the many who stimulated discussion at the symposium. We also appreciate the excellent work of the ASA Headquarters staff in the publication process.

Organizing Committee

W.L. Hargrove
A.L. Black
J.V. Mannering

Editorial Committee

W.L. Hargrove, co-chair
B.G. Ellis, co-chair
T. Cavalieri
R.C. Johnson
R.J. Reginato

CONTRIBUTORS

R. W. Andrews	Editorial Research Assistant, Rodale Research Center, Kutztown, Pennsylvania
A. L. Black	USDA-ARS, Northern Great Plains Research Lab., Mandan, North Dakota
R. L. Blevins	Professor of Agronomy, Department of Agronomy, University of Kentucky, Lexington, Kentucky
S. J. Corak	Graduate Research Assistant, Department of Agronomy, University of Kentucky, Lexington, Kentucky
J. W. Doran	Soil Scientist, USDA-ARS, University of Nebraska, Lincoln, Nebraska
W. W. Frye	Professor of Agronomy, Department of Agronomy, University of Kentucky, Lexington, Kentucky
G. J. Gascho	Professor, Department of Agronomy, Coastal Plains Experiment Station, University of Georgia, Tifton, Georgia
A. D. Halvorson	Soil Scientist, USDA-ARS, Akron, Colorado
W. L. Hargrove	Associate Professor, Department of Agronomy, Georgia Agricultural Experiment Station, University of Georgia, Griffin, Georgia
O. B. Hesterman	Assistant Professor, Crop and Soil Sciences Department, Michigan State University, East Lansing, Michigan
J. E. Hook	Associate Professor, Agronomy Department, Coastal Plain Experiment Station, University of Georgia, Tifton, Georgia
R. R. Janke	Agronomy Coordinator, Rodale Research Center, Kutztown, Pennsylvania
D. L. Karlen	Soil Scientist, USDA-ARS, Department of Agronomy, Iowa State University, Ames, Iowa
G. W. Langdale	USDA-ARS, Southern Piedmont Conservation Research Center, Watkinsville, Georgia
J. V. Mannering	Professor, Department of Agronomy, Purdue University, West Lafayette, Indiana
D. B. Mengel	Professor of Agronomy, Department of Agronomy, Purdue University, West Lafayette, Indiana
R. I. Papendick	Soil Scientist, USDA-ARS, Land Management and Water Conservation, Washington State University, Pullman, Washington
S. E. Peters	Research Agronomist, Rodale Research Center, Kutztown, Pennsylvania
Francis J. Pierce	Assistant Professor, Department of Crop and Soil Sciences, Michigan State University, East Lansing, Michigan

J. F. Power	Research Leader, USDA-ARS, University of Nebraska, Lincoln, Nebraska
J. K. Radke	Research Soil Scientist, USDA-ARS, Rodale Research Center, Kutztown, Pennsylvania
Charles W. Rice	Assistant Professor, Department of Agronomy, Kansas State University, Manhattan, Kansas
E. J. Sadler	Soil Scientist, USDA-ARS, Coastal Plains Soil and Water Conservation Research Center, Florence, South Carolina
James S. Schepers	Soil Scientist, USDA-ARS, University of Nebraska, Lincoln, Nebraska
M. S. Smith	Professor of Agronomy, Department of Agronomy, University of Kentucky, Lexington, Kentucky
R. E. Sojka	Soil Scientist, USDA-ARS, Snake River Soil and Water Management Research Unit, Kimberly, Idaho
P. W. Unger	Soil Scientist, USDA-ARS, Conservation and Production Research Lab., Bushland, Texas
J. J. Varco	Assistant Professor of Agronomy, Department of Agronomy, Mississippi State University, Mississippi State
M. G. Wagger	Assistant Professor, Department of Crop Science, North Carolina State University, Raleigh, North Carolina

Conversion Factors for SI and non-SI Units

Conversion Factors for SI and non-SI Units

To convert Column 1 into Column 2, multiply by	Column 1 SI Unit	Column 2 non-SI Unit	To convert Column 2 into Column 1 multiply by
		Length	
0.621	kilometer, km (10^3 m)	mile, mi	1.609
1.094	meter, m	yard, yd	0.914
3.28	meter, m	foot, ft	0.304
1.0	micrometer, μm (10^{-6} m)	micron, μ	1.0
3.94×10^{-2}	millimeter, mm (10^{-3} m)	inch, in	25.4
10	nanometer, nm (10^{-9} m)	Angstrom, Å	0.1
		Area	
2.47	hectare, ha	acre	0.405
247	square kilometer, km^2 (10^3 m)^2	acre	4.05×10^{-3}
0.386	square kilometer, km^2 (10^3 m)^2	square mile, mi^2	2.590
2.47×10^{-4}	square meter, m^2	acre	4.05×10^3
10.76	square meter, m^2	square foot, ft^2	9.29×10^{-2}
1.55×10^{-3}	square millimeter, mm^2 (10^{-6} m)^2	square inch, in^2	645
		Volume	
9.73×10^{-3}	cubic meter, m^3	acre-inch	102.8
35.3	cubic meter, m^3	cubic foot, ft^3	2.83×10^{-2}
6.10×10^4	cubic meter, m^3	cubic inch, in^3	1.64×10^{-5}
2.84×10^{-2}	liter, L (10^{-3} m^3)	bushel, bu	35.24
1.057	liter, L (10^{-3} m^3)	quart (liquid), qt	0.946
3.53×10^{-2}	liter, L (10^{-3} m^3)	cubic foot, ft^3	28.3
0.265	liter, L (10^{-3} m^3)	gallon	3.78
33.78	liter, L (10^{-3} m^3)	ounce (fluid), oz	2.96×10^{-2}
2.11	liter, L (10^{-3} m^3)	pint (fluid), pt	0.473

	Mass		
2.20×10^{-3}	gram, g (10^{-3} kg)	pound, lb	454
3.52×10^{-2}	gram, g (10^{-3} kg)	ounce (avdp), oz	28.4
2.205	kilogram, kg	pound, lb	0.454
0.01	kilogram, kg	quintal (metric), q	100
1.10×10^{-3}	kilogram, kg	ton (2000 lb), ton	907
1.102	megagram, Mg (tonne)	ton (U.S.), ton	0.907
1.102	tonne, t	ton (U.S.), ton	0.907
	Yield and Rate		
0.893	kilogram per hectare, kg ha^{-1}	pound per acre, lb $acre^{-1}$	1.12
7.77×10^{-2}	kilogram per cubic meter, kg m^{-3}	pound per bushel, lb bu^{-1}	12.87
1.49×10^{-2}	kilogram per hectare, kg ha^{-1}	bushel per acre, 60 lb	67.19
1.59×10^{-2}	kilogram per hectare, kg ha^{-1}	bushel per acre, 56 lb	62.71
1.86×10^{-2}	kilogram per hectare, kg ha^{-1}	bushel per acre, 48 lb	53.75
0.107	liter per hectare, L ha^{-1}	gallon per acre	9.35
893	tonnes per hectare, t ha^{-1}	pound per acre, lb $acre^{-1}$	1.12×10^{-3}
893	megagram per hectare, Mg ha^{-1}	pound per acre, lb $acre^{-1}$	1.12×10^{-3}
0.446	megagram per hectare, Mg ha^{-1}	ton (2000 lb) per acre, ton $acre^{-1}$	2.24
2.24	meter per second, m s^{-1}	mile per hour	0.447
	Specific Surface		
10	square meter per kilogram, m^2 kg^{-1}	square centimeter per gram, cm^2 g^{-1}	0.1
1 000	square meter per kilogram, m^2 kg^{-1}	square millimeter per gram, mm^2 g^{-1}	0.001
	Pressure		
9.90	megapascal, MPa (10^6 Pa)	atmosphere	0.101
10	megapascal, MPa (10^6 Pa)	bar	0.1
1.00	megagram per cubic meter, Mg m^{-3}	gram per cubic centimeter, g cm^{-3}	1.00
2.09×10^{-2}	pascal, Pa	pound per square foot, lb ft^{-2}	47.9
1.45×10^{-4}	pascal, Pa	pound per square inch, lb in^{-2}	6.90×10^3

continued on next page

Conversion Factors for SI and non-SI Units

To convert Column 1 into Column 2, multiply by	Column 1 SI Unit	Column 2 non-SI Unit	To convert Column 2 into Column 1 multiply by
		Temperature	
1.00 (K − 273)	Kelvin, K	Celsius, °C	1.00 (°C + 273)
(9/5 °C) + 32	Celsius, °C	Fahrenheit, °F	5/9 (°F −32)
		Energy, Work, Quantity of Heat	
9.52×10^{-4}	joule, J	British thermal unit, Btu	1.05×10^{3}
0.239	joule, J	calorie, cal	4.19
10^{7}	joule, J	erg	10^{-7}
0.735	joule, J	foot-pound	1.36
2.387×10^{-5}	joule per square meter, J m^{-2}	calorie per square centimeter (langley)	4.19×10^{4}
10^{5}	newton, N	dyne	10^{-5}
1.43×10^{-3}	watt per square meter, W m^{-2}	calorie per square centimeter minute (irradiance), cal cm^{-2} min^{-1}	698
		Transpiration and Photosynthesis	
3.60×10^{-2}	milligram per square meter second, mg m^{-2} s^{-1}	gram per square decimeter hour, g dm^{-2} h^{-1}	27.8
5.56×10^{-3}	milligram (H_2O) per square meter second, mg m^{-2} s^{-1}	micromole (H_2O) per square centimeter second, μmol cm^{-2} s^{-1}	180
10^{-4}	milligram per square meter second, mg m^{-2} s^{-1}	milligram per square centimeter second, mg cm^{-2} s^{-1}	10^{4}
35.97	milligram per square meter second, mg m^{-2} s^{-1}	milligram per square decimeter hour, mg dm^{-2} h^{-1}	2.78×10^{-2}
		Plane Angle	
57.3	radian, rad	degrees (angle), °	1.75×10^{-2}

	Electrical Conductivity, Electricity, and Magnetism		
10	siemen per meter, $S\ m^{-1}$	millimho per centimeter, $mmho\ cm^{-1}$	0.1
10^4	tesla, T	gauss, G	10^{-4}
	Water Measurement		
9.73×10^{-3}	cubic meter, m^3	acre-inches, acre-in	102.8
9.81×10^{-3}	cubic meter per hour, $m^3\ h^{-1}$	cubic feet per second, $ft^3\ s^{-1}$	101.9
4.40	cubic meter per hour, $m^3\ h^{-1}$	U.S. gallons per minute, $gal\ min^{-1}$	0.227
8.11	hectare-meters, ha-m	acre-feet, acre-ft	0.123
97.28	hectare-meters, ha-m	acre-inches, acre-in	1.03×10^{-2}
8.1×10^{-2}	hectare-centimeters, ha-cm	acre-feet, acre-ft	12.33
	Concentrations		
1	centimole per kilogram, $cmol\ kg^{-1}$ (ion exchange capacity)	milliequivalents per 100 grams, meq $100\ g^{-1}$	1
0.1	gram per kilogram, $g\ kg^{-1}$	percent, %	10
1	milligram per kilogram, $mg\ kg^{-1}$	parts per million, ppm	1
	Radioactivity		
2.7×10^{-11}	bequerel, Bq	curie, Ci	3.7×10^{10}
2.7×10^{-2}	bequerel per kilogram, $Bq\ kg^{-1}$	picocurie per gram, $pCi\ g^{-1}$	37
100	gray, Gy (absorbed dose)	rad, rd	0.01
100	sievert, Sv (equivalent dose)	rem (roentgen equivalent man)	0.01
	Plant Nutrient Conversion		
	Elemental	***Oxide***	
2.29	P	P_2O_5	0.437
1.20	K	K_2O	0.830
1.39	Ca	CaO	0.715
1.66	Mg	MgO	0.602

5 May 1988

1 Cropping Strategies for Efficient Use of Water and Nitrogen: Introduction[1]

W. L. Hargrove

Georgia Agricultural Experiment Station
Griffin, Georgia

A. L. Black

Northern Great Plains Research Laboratory
Mandan, North Dakota

J. V. Mannering

Purdue University
West Lafayette, Indiana

Currently, crop production in the USA can be characterized by (i) low commodity prices, (ii) relative high input costs, and (iii) increasing concern over the influence of modern agriculture on environmental quality. Recent trends in corn (*Zea mays* L.) and soybean (*Glycine max* L. Merr.) prices are shown in Fig. 1-1. Since 1983, prices in the USA have declined for these commodities which represent several other important grain and oil seed crops. This decline is related to a surplus in world supply of most crop commodities and a decrease in U.S. exports (Runge, 1986). The decline in U.S. exports was related, in turn, to an overvalued dollar, inflating prices paid by foreign buyers (Runge, 1986). The currently deflated value of the dollar, however, has not nor is it expected to help agricultural exports because of large grain surpluses.

While commodity prices have been dropping, production costs have been increasing. Average production costs for corn and soybean grown in the Corn Belt over the past 10 yr are shown in Fig. 1-2. These values are generally less than those from other regions of the country (Robinson, 1986), but trends are similar. Since 1979, input costs have increased significantly for several reasons. Crop production in the USA is highly mechanized, requiring little labor but dependent instead on large and specialized equipment, relatively

[1] Contribution from the Georgia Agric. Exp. Stn., USDA-ARS, and Purdue University. Supported by state, Hatch, and federal funds allocated to the Georgia Agric. Exp. Stn., USDA-ARS, and Purdue Univ.

Cropping Strategies for Efficient Use of Water and Nitrogen, Special Publication no. 51.

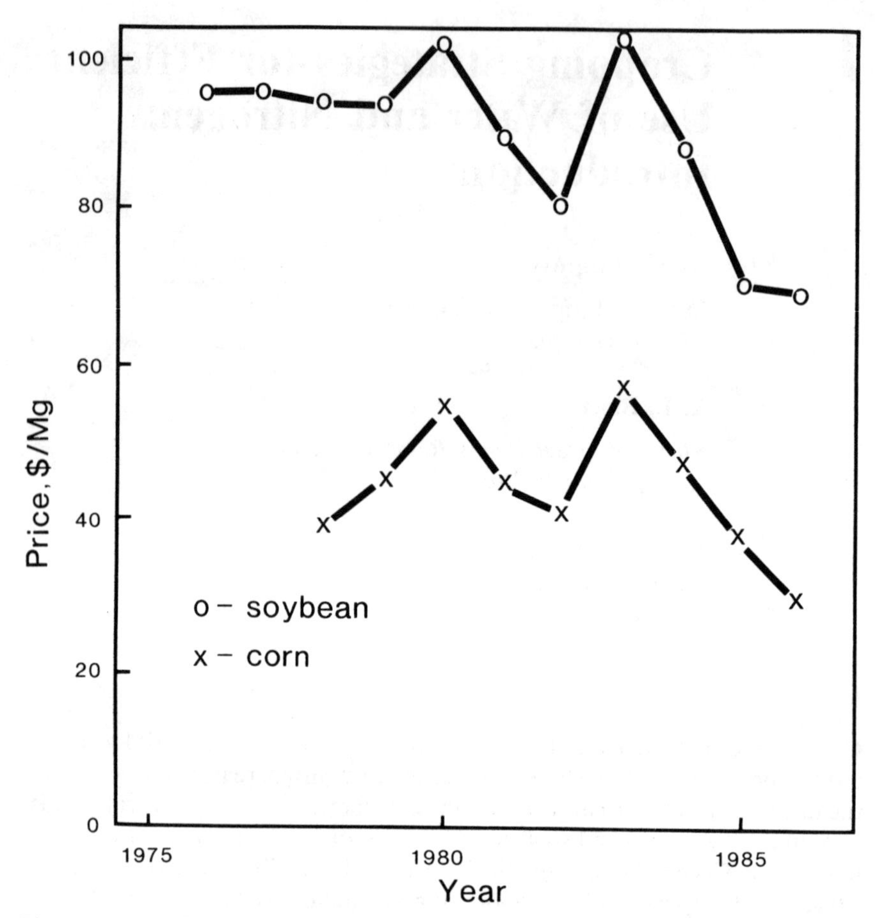

Fig. 1–1. Recent trends in corn and soybean price. From *Georgia Agricultural Facts* (1986 ed. Georgia Agric. Statistics, Athens.)

large fossil fuel energy inputs, and intensive fertilizer and pesticide use. Furthermore, over the past few years, producers have incurred a high rate of indebtedness for capital and operating expenditures causing interest paid on borrowed money to become a major expense. In addition, land values have declined dramatically, thus greatly decreasing the landowner's worth and decreasing his opportunity to borrow money.

The recent decline in prices coupled with the increase in production costs have resulted in a narrow margin of profitability for most crops. Percent return on investment for corn, soybean, and wheat (*Triticum aestivum* L.) production in various regions of the country is shown in Table 1–1. At the 1985 harvest prices, only soybean in the Lake States, Corn Belt, and Central Plains provides producers with an investment return equal to their 20-yr average rate of 4.33% (Ott et al., 1986).

In addition, current crop production has become less and less diversified. Continuous monocropping with little or no crop rotation is questionable

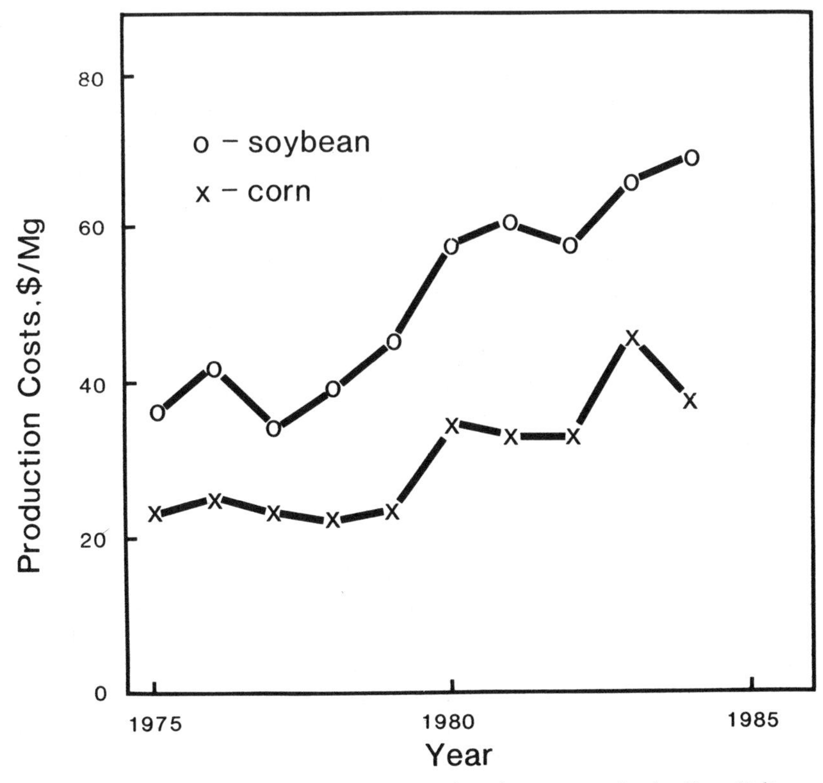

Fig. 1-2. Average production costs for corn and soybean grown in the Corn Belt over the past 10 yr. From Robinson (1986).

ecologically, as well as economically. The lack of diversification leaves the producer vulnerable to disaster at the hands of crop pests (weeds, diseases, and insects) and/or economics (low price and high costs). Continuous monoculture tends to favor individual weeds, diseases, and/or insects. For example, Buchanan et al. (1975) in a 3-yr study, showed that higher populations of common cocklebur (*Xanthium pensylvanicum*) were associated more with soybean than with corn regardless of weed control methods. Crop rota-

Table 1-1. Return on investment (%) for corn, soybean, and wheat in 1985. From Ott et al. (1986).

Region	Corn†	Soybean‡	Wheat§
Lake States	NR¶	5.3	NR
Corn Belt	2.7	4.6	NR
Central Plains	NR	7.2	2.4
Southeast	NR	NR	NR
Delta	--	1.5	NR

† Based on a price of $34.40 Mg^{-1} ($2.16/bu).
‡ Based on a price of $72.90 Mg^{-1} ($4.90/bu).
§ Based on a price of $42.71 Mg^{-1} ($2.87/bu).
¶ Negative return.

tion, therefore, is valuable in managing crop pests as well as in minimizing economic risks.

Superimposed on these economic factors are growing concerns over surface and groundwater contamination by fertilizers and pesticides (Baker and Laflen, 1983; Hallberg, 1987). According to Hallberg (1987), research worldwide has shown increases in NO_3-N in groundwater concurrent with major increases in N fertilization, and many shallow water supplies now exceed the 10 mg of NO_3-N L^{-1} drinking water limit. In the continental USA, 39 states have confirmed reports of significant surface and/or groundwater contamination by nutrients and/or pesticides (USDA, 1987). In addition, excessive soil erosion rates are undermining our ability to produce acceptable yields and also adversely impact water quality (ASAE, 1987). More than 70 million ha of cropland are eroding at rates greater than the soil loss tolerance, the rate at which sustained production is possible, and 23% of cropland is eroding at twice the tolerable rate (USDA, 1987).

In the near future, several factors could exacerbate an already strained situation in crop production. One such factor is the strong likelihood of a new U.S. energy crisis in the early to mid-1990s, prompted by increasing U.S. dependence on importation of foreign oil (Hirsch, 1987). Due to the weakening U.S. oil and gas industry, dependence on foreign oil is expected to rise from 32% in 1983 to 50 to 70% by 1995 (Hirsch, 1987). Energy availability and price strongly influence N-fertilizer price, as well as direct costs for tractor fuel and fuel for pumping water for irrigation. The adverse effect of a significant price increase in energy on profitability in farming is therefore obvious.

Also, the profusion of pending water-quality legislation to lessen the off-site impact of intensive crop production will increasingly constrain pesticide and fertilizer use and narrow soil and crop management options. In a recent issue of *Agronomy News* (Nipp, 1987), approximately 35 pieces of pending legislation affecting agricultural research were summarized. Of these, about 45% were related to environmental quality issues, particularly NO_3-N and pesticide contamination of groundwater.

Finally, it appears that low grain prices caused by world surpluses of grains and high input costs will continue to result in a narrow margin of profitability for most grain crops. Although many other factors, such as trade and domestic agricultural policies, monetary exchange rates, world economic growth rates, inflation, and recession also influence agricultural profitability, the trends are for the competetiveness of U.S. agriculture to continue to erode (Robinson, 1986).

The narrow margin of profitability for most crops, the overall depressed farm economy, predictions of sharp increases in fossil fuel costs, and growing concerns over deterioration of soil and water resources suggest that current production technologies are not sustainable. To counter these factors, agronomists must alter existing crop production systems. The challenge facing agronomists is to develop new strategies for crop production within a framework of increasing environmental and economic constraints. Development of sustainable production technologies will require close scrutiny of necessary inputs and renewed efforts to improve crop production efficiency.

Water and N are two of the most critical inputs in crop production. For nonirrigated conditions, water availability dictates crop production more than any other factor. Nitrogen is generally the most limiting plant nutrient in soils worldwide. Therefore, relatively large amounts of fertilizer-N are required for acceptable crop yields. Since water and N are relatively large inputs and strongly influence productivity and water is the carrier of excess NO_3-N to groundwater, their use by crops and ultimate fate are closely intertwined. It therefore seems appropriate to focus on efficient water and N use as a first step in development of sustainable production technologies.

The objectives of this publication are to: (i) develop concepts of crop water and N use within a framework of increasing environmental and economic constraints, and (ii) propose strategies for improved crop water and N use within these constraints. Crop management practices such as rotation, planting geometries, multiple cropping, crop residue management, and cover crops are discussed in light of water and N-use efficiency. It is the intent of the authors that this serve as a research initiative for development of sustainable crop production systems that are both profitable and environmentally sound.

REFERENCES

American Society of Agricultural Engineers. 1987. Erosion and soil productivity. *In* Proc. Natl. Symp. Erosion Soil Product, New Orleans. 10-11 Dec., 1984. ASAE, St. Joseph, MI.

Baker, J.L., and J.M. Laflen. 1983. Water quality consequences of conservation tillage. J. Soil Water Conserv. 38:186-193.

Buchanan, G.A., C.S. Hoveland, V.L. Brown, and R.H. Wade. 1975. Weed population shifts influenced by crop rotations and weed control programs. Proc. South. Weed Sci. Soc. 28:60-71.

Hallberg, G.R. 1987. Agricultural chemicals in groundwater: Extent and implications. Am. J. Alter. Agric. 2:3-15.

Hirsch, R.L. 1987. Impending United States energy crisis. Science 235:1467-1473.

Nipp, T.L. 1987. Congressional update. p. 1 *In* Agronomy news. (Sept.). ASA, Madison, WI.

Ott, S.L., J.R. Allison, and G.A. Shumaker. 1986. The competitive position of the southeast in national and international markets for corn, soybeans, and wheat. J. Agribusiness 4:27-33.

Robinson, B.H. 1986. Can farmers in the Southeast compete in world markets? J. Agribusiness 4:16-26.

Runge, C.F. 1986. Obstacles and opportunities confronting U.S. agriculture in world markets. J. Agribusiness 4:3-5.

U.S. Department of Agriculture. 1987. The second RCA appraisal. July-August Review Draft. U.S. Gov. Print. Office, Washington, DC.

2 Multiple Cropping for Efficient Use of Water and Nitrogen[1]

J. E. Hook and G. J. Gascho
University of Georgia
Tifton, Georgia

Alfalfa (*Medicago sativa* L.) and timothy (*Phleum pratense* L.) cover hay fields in the Northeast; wheat (*Triticum aestivum* L.) is planted in the fall after soybean [*Glycine max* (L.) Merr.] harvest in the Southeast; corn (*Zea mays* L.) is planted between rows of young pecan (*Carya illinoensis*) trees in Louisiana; mung bean (*Phaseolus aureus* L.) climbs the drying stalks of corn in central Africa; sweet potato (*Ipomoea batatus* Lam.) is sown after a crop of tomato (*Lycopersicon esculentum* L.) transplants have been pulled in the Southeast. What each of these farming systems has in common is the attempt to produce more from a parcel of land than would be obtained by one crop alone. Multiple cropping has been practiced since antiquity; however, the possibility that multiple cropping may increase productivity of lands and make more efficient use of resources has led to a renewed interest in multiple cropping. The possibility of distributing annual fixed costs over more than one crop, changing patterns of cash flow, and reducing risks of total crop failure are some of the economic incentives that have encouraged multiple cropping. Possibilities for reduction in soil erosion, for retention of soluble nutrients, and for minimizing labor needed for land preparation favor planting a second crop, if only as a cover crop.

This chapter examines, largely from a theoretical standpoint, strategies for improvements in crop water and N use by using multiple cropping. The combinations of possible multiple-cropping systems, each with its unique problems and advantages, are too numerous to discuss here, but a general definition of systems is given. Similarly, the definitions of water and N-use efficiency vary greatly depending on the purposes for examining efficiencies. An ecologically based definition is given here and used throughout. The focus in this chapter is on water and N-use efficiencies obtained through multiple cropping that are above and beyond those that may be obtained within a single crop. Other chapters in this book consider water and N use specifically within crops. A previous symposium (Papendick et al., 1976) and a recent book (Francis, 1986) provide an examination of multiple cropping from many

[1]Contribution of the Dep. of Agronomy, Univ. of Georgia, Coastal Plain Stn., Tifton, GA 31793.

Cropping Strategies for Efficient Use of Water and Nitrogen, Special Publication no. 51.

perspectives. No attempt is made to justify multiple cropping, as this must come from a complete analysis of all possible competing crop systems. A thorough discussion of the socio-economic considerations of multiple cropping is provided by Hildebrand (1976).

MULTIPLE CROPPING

Multiple cropping systems can be subdivided into sequential and intercrop systems. In sequential multiple cropping, more than one crop is grown per year on the same land, but one crop follows another. With intercropping, the two crops share the same land for all or part of their growing seasons.

Intercropping, as used in developing countries, can resemble family gardening with a wide variety of vegetable and ornamental plants on a small parcel of land, or it can cover more extensive areas using only two species, such as a legume with corn. Although many intercropping systems lend themselves to hand harvesting and are unsuitable where labor is expensive, there are several important intercropping practices that allow machine harvesting. The most common are mixed forage stands used for hay. Although managed as a single crop, mixed stands use plants with differing growth habits, nutrient requirements, temperature responses, or forage quality to achieve increased efficiency of land, water, and nutrient use. Other intercropping practices include the planting of grain into established orchards, overseeding forages with winter or summer annuals, and the conservation practices of planting corn or sorghum into self-seeding legumes or other living forages.

Sequential multiple-cropping systems are more common in the USA than intercropping. They can be established wherever the frost-free season is longer than the single crop-growing season. Commonly, the second crop is a winter cover or winter grain crop planted after the summer crop is harvested. Vegetable crops, which may have a growing season as short as 30 d, allow a wider selection of sequential multiple-cropping options. In regions where frost occurs infrequently or for only short durations, numerous crop selections are possible to make use of the land during hot summers and cool winters.

UTILIZATION EFFICIENCIES

Efficiency is a measure of output per unit input. Because we rarely consider all the outputs or all the inputs involved in crop production, efficiency is somewhat narrowly defined. The definitions chosen, however, must encompass all practical inputs and outputs of the system being studied. For example, CO_2 assimilation rate vs. leaf transpiration rate, the A/E ratio, is a water-use index that narrowly examines the efficiency of the plant leaves fixing C while losing water vapor (Frank et al., 1987). The ratio may be useful in physiological or breeding efforts, but is too narrow a definition of water-use efficiency to be applied in all cases. Total dry matter or grain production per total season evapotranspiration (ET) describes ET-based water-use efficiency (ET-WUE). This index is used commonly to examine crop response to various water parameters. Hillel and Guron (1973) found a linear relation between corn grain or silage and seasonal ET, and noted that the relationship differed slightly from season to season in the Israeli Negev region. Over 3 yr, corn yield (Y in kg ha^{-1}) was related to seasonal ET (mm) by the equation:

$$Y = (41.0 \cdot ET) - 7179. \quad [1]$$

Similar 3-yr average responses of corn grain to ET for Bushland, TX (Musick and Dusek, 1980), Gainesville, FL (Hammond et al., 1981), and Akron, CO (Hanks et al., 1969) were described, respectively, by the following equations:

$$Y = (24.1 \cdot ET) - 8332 \quad [2]$$

$$Y = (66.1 \cdot ET) - 22\,900 \quad [3]$$

$$Y = (20.8 \cdot ET) - 1640. \quad [4]$$

The variation in these equations suggests considerable variation in the ET-WUE relationship by climate or other factors. The negative intercept indicates that a minimum threshold of ET is required before any measurable yield occurs. At least part of this threshold is due to the soil evaporation that occurs before newly emerged plants begin accumulating measurable dry matter. Using transpiration as the input of the efficiency equation gives a transpiration-based water-use efficiency index (T-WUE). A linear relationship exists between biomass and transpiration that has an intercept more nearly equal to zero (Stegman et al., 1980). Full-season transpiration is more difficult to measure than full-season ET, however, and this T-WUE index is used less frequently than ET-WUE. The response of dry matter production to water lost as ET or transpiration has been observed to be linear for most other crops (Stegman et al., 1980; Doorenbos and Kassam, 1979). This makes ET-WUE particularly useful for predicting which crops will be affected most severely by water deficits in various locations for which the ET-WUE relationship is known.

The dry matter/ET ratio considers only the water that was used as transpiration and evaporation, but it ignores what was actually applied or available. Irrigation water-use efficiency (I-WUE) bases efficiency on the total irrigation water applied (Musick and Dusek, 1980). Except in arid climates where irrigation may be related to seasonal ET consumption, the I-WUE will be strongly dependent upon the rainfall which contributed to production of the biomass. Often, only the biomass production above that produced in rainfed conditions is considered in I-WUE computations. Even so, comparisons among computed I-WUE values can only be made for identical rainfall and other conditions.

For N, similar indices of efficiency are used (Zweifel et al., 1987). Nitrogen fertilizer-use efficiency is often measured as grain removed from a field vs. N in fertilizer added just before or during the growing period of that crop. This narrow definition ignores N input from other sources and N cycling within the boundaries of the system of interest. Each of these definitions is useful in certain situations but places restrictions on which part of the input or output is included. Although this is often necessary, each of these efficiency definitions excludes some of the inputs of the water and N upon which multiple cropping draws to gain its efficiency.

For multiple cropping, other definitions are more appropriate. Just as the factory is the manufacturer's basic unit for production, the land is the farmer's basic unit of production. We often examine the efficiency with which a farmer produces a monocrop, considering only the resources available to that crop during its growing season. Multiple cropping, however, encourages us to examine how efficiently the farmer uses that basic production unit, the land, throughout the entire year. The boundaries of the production unit are the edges of the fields and the depth of soil from which the resources can be extracted.

For the intercropping systems often found in the tropics, Harwood and others (IRRI, 1974) proposed the land equivalence ratio (LER) as a measure of efficiency. The LER compares production of two or more crops in a multiple-cropping sequence, especially an intercropping system, to the production of the same crops in monoculture. For each crop in the intercropping system, for example mung bean growing on corn, a relative yield of each species is computed:

$$\text{Relative bean yield} = \text{bean yield (intercrop)/bean yield (monocrop)},$$

and

$$\text{Relative corn yield} = \text{corn yield (intercrop)/corn yield (monocrop)},$$

The LER, then, is relative bean yield plus relative corn yield. In a two-species intercrop such as this, each of the species needs to produce only half as much yield in intercrop as it would in the monocrop to achieve an LER $= 1.0$. An LER > 1.0 suggests that less land would be required to produce the same biomass or food in intercrop as in monoculture. Many reports of high LER have been noted suggesting a land-use advantage to the cropping sequences studied (Trenbath, 1974, 1976; IRRI, 1974; Wiley, 1985).

Because multiple crops effectively expand the growing season, the time during which crops occupy the land should be considered. Hiebsch and McCollum (1987) proposed the area $\times$ time equivalence ratio (ATER) which considered not only the amount of land occupied by the intercrop but also the amount of time the entire intercrop ties-up the land as compared with monocrops. By computing production per unit of land per unit of time, the

ATER index removed much of the speculation about how intercrops could outproduce monocrops. In the corn-mung bean intercropping example, the bean has a shorter growing season than the corn. If the bean had been planted in monoculture, the land would have been available for planting another crop earlier than with the intercrop. The LER tends to overestimate the intercropping efficiency advantage by ignoring the production of the crop that could have followed mung bean in a monoculture. Hiebsch and McCollum (1987) noted that in intercropping where one crop supplied a limiting growth factor to its partner, there was a true gain in efficiency with the intercropping over the monoculture. Typically, this occurred in legume-nonlegume mixtures where soil and fertilizer N was low.

The difficulty with the ATER index for multiple cropping is that, similar to the ET-WUE index, it examines only the land and time which were used during the production of a crop and not the total time and land which potentially could have been used. For practical purposes, the time boundary for examining efficiencies in multiple cropping should be 1 yr. A strong argument could be made for considering one crop rotation cycle as the time boundary. Our experience, however, suggests that rotations continue to change and evolve as market conditions change, and it is often difficult to decide when a rotation has ended. Also, because economic analyses are most commonly computed annually, there is additional benefit for the use of 1 yr as a base for determining efficiencies.

While the LER and ATER indices have helped to understand production efficiencies in multiple-cropping systems, these methods consider only land and time resources. To include the water and N resources, we refer to the concept of resource-use efficiency (RUE) used by Trenbath (1986) in analyzing light interception by intercrops. For inputs, RUE considers all the components of a resource available to a crop or to a cropping system within the area and time boundaries. For outputs, biomass, grain, protein, or other unit of harvested crop could be used. However, comparison of RUE of a sweet potato-potato-peanut-sorghum [*Ipomoea batatus* (L.) Lam,-*Solanum tuberosum* L.-*Arachis hypogaea* L.-*Sorghum bicolor* (L.) Moench] rotation with a corn-snap bean-winter rye-squash-soybean (*Zea mays* L.-*Phaseolus vulgaris* L.-*Secale cereale* L.-*Cucurbita maxima* L.-*Glycine max* L. Merr.) rotation demands a more common output unit than biomass. As argued by Hildebrand (1976), the profitability of the two systems per unit area is the ultimate arbiter. A realistic and useful way to compare such diverse rotations is by the value of the total product, in spite of the variable nature of prices. Thus, the output for RUE computations must be the marketable portion of the crops in the system.

This definition of efficiency is more comprehensive than many, and the computation can be as difficult in practice as the measurement of all the components of the resource inputs. However, the RUE concept stresses the importance of the entire production unit, the land, over the entire crop rotation cycle. Fertilizer losses that occur after the growing season are true losses to that production unit, even though they have no direct effect on efficiency during the production of the crop to which they were applied. By assigning a value to other outputs besides the marketable portion of the crop, RUE can be expanded to become an ecological system definition. Recent efforts to quantify effects of erosion on productivity are attempts to assign a value to the soil output (erosion losses) from a cropping system. Resource-use efficiency is a comprehensive index valuable in multiple-cropping systems.

Trenbath (1986) recognized that two components of resource-use efficiency were capture and conversion. *Capture efficiency* refers to the crop's ability to gather as much of the available resource as possible. In the case of solar radiation resource, efficient capturers develop canopy architectures, leaf densities, and colorations that intercept much of the photosynthetically active radiation preventing it from reaching the soil or reflecting into the sky. *Conversion efficiency* refers to the ability to produce biomass by using the captured and stored resource. Efficient converters of the captured light fix and store C while minimizing loss of the light energy through heat loss. The T-WUE index mentioned previously describes conversion efficiency of water. However, both capture and conversion are necessary for efficient resource use.

Before further refining the RUE concept, a few examples show why it is useful. Nitrogen and water are unevenly distributed in time and space. Mineralized N may be most readily available in late spring when supplies of easily decomposed residues, favorable moisture conditions, and increasing soil temperatures result in rapid decomposition, mineralization, and nitrification. A sorghum crop actively taking up N at that time may capture much of the mineralized N. If that N is effectively converted into protein which is harvested, the use (capture and conversion) efficiency of that resource (mineralized N) will be high. A winter wheat crop, on the other hand, would be unable to capture this source of N because its uptake period would precede the rapid mineralization period. It would have a lower RUE for the mineralized N. The soil water resource may be more readily available in heavy-textured subsoil layers than in a sandy topsoil. A peanut crop, noted for its deep-rooting habit, would be able to capture that water. If it converts (transpires) that water to meet transpiration demands during critical growth periods the use efficiency of that resource (subsoil water) would be high. A shallow rooted lettuce (*Lactuca sativa* L.) would be unable to capture this deep water. Its RUE of the subsoil water use would be low even if its conversion efficiency (T-WUE) were high. Plants differ in their abilities to absorb these unevenly distributed environmental resources and in their abilities to convert them to dry matter. Thus, RUE provides a comprehensive definition from which we can consider these components as available to the crop.

SOURCES OF WATER AND NITROGEN-USE EFFICIENCY

In general, we can assume that practices that improve water or N RUE within a single crop system will improve efficiency in the total multiple crop system. Because discussion of these has been the focus of other reports in this book, we consider those conditions that improve efficiency over several crops rather than within individual crops of the multiple crop sequence. To examine water and N RUE, the boundaries and inputs must be further defined. Because water and N are obtained through the root zone, the physical lower boundary for computing resource availability is the potential rooting depth of the deepest rooted crop considered in the intercrop or multiple crop sequence. For many situations, this depth is the depth to a water table, bedrock, or other impeding layers. This lower boundary is used in order to examine capture efficiencies. When water or N moves below the shallow roots of a lettuce crop, it leads to a low capture efficiency. However, if that water or N is subsequently captured by the corn crop which follows, the capture efficiency of the multiple-crop sequence can still be high.

Water inputs include all water that reaches the plant canopy or soil surface as rainfall or irrigation during the year. All N that reaches the field in rainfall, fertilizers, manures, and other amendments are considered. Atmospheric N presents a problem. Even the best legume-*Rhizobium* associations capture little of the total atmospheric N available to them. For simplification, we exclude atmospheric N as an input until that N is fixed.

WATER RESOURCE-USE EFFICIENCY IN MULTIPLE CROPPING

Increased Use of Annual Rainfall

Annual rainfall that occurs after the yield portion of the crop is complete and before the next year's crop is planted refills depleted soil water storage or leaves the boundaries of the cropping system as runoff or deep percolation. A second crop planted during this off season will directly use some of the water that otherwise may leave the soil before a crop is planted in the next year.

A monoculture of soybean in the Coastal Plain of Georgia may produce 2700 kg ha^{-1} of bean during its May through September growing season. Assuming they dewater to the 1.2-m soil depth before harvest and that they receive 550 mm of rain, the average rainfall for this period, the water RUE of the soybean-growing season would be 3.6 kg ha^{-1} mm^{-1}. Double-cropping wheat during the winter-spring period requires a shifting of soybean planting and harvest dates by 30 d, but yield of the soybean is not necessarily reduced in this region. Doublecropping uses the land during the entire year and relies upon the 1180-mm average annual rainfall. If the same soybean yield is achieved, and 2400 kg ha^{-1} of wheat is produced, the water RUE of the double crop is 4.3 kg ha^{-1} mm^{-1}. If the water RUE of the soybean monocrop were computed over the entire year, it would be only 2.3 kg ha^{-1} mm^{-1}. It is easy to see that, even with a slight reduction in soybean yield with doublecropping, the water RUE for the year can still be greater than with monocropping.

By increasing the use of this annual rainfall, both capture efficiency and conversion efficiency of the water can be improved over the year. The vegetative cover provided by the second crop can improve infiltration during the off season resulting in more water utilized by that crop or by later main crops. If the second crop is harvested, off-season rainfall is effectively converted to biomass.

Increased Use of Irrigation Water

When vegetables, transplants, ornamentals, and other high-value crops are grown, irrigation is often continued until harvest to assure high quality of the product. At harvest, soil water storage may be nearly full. By immediately following the high-value crop with a second crop this water can be utilized, and the total conversion efficiency of the irrigation resource is improved.

Miscellaneous Improvements in Water Capture or Use Efficiency

Indirect benefits may accrue from planting secondary crops during the off-season. Improved soil structure associated with a growing root system and improved ground cover can carry over to improve infiltration and aid water-capture efficiency of the main crop (Jones et al., 1969; Moody et al., 1963). Beale et al. (1955) pointed out many years ago that rye and vetch (*Vicia* spp.) as winter cover led to improved soil aggregation, and increased organic matter as compared to a conventionally plowed check treatment. The degree of soil aggregation improved during 10 yr of winter cover management, and an average of 78 mm of additional infiltration per year was observed. Thus, even when a second crop is not harvested, e.g., a winter cover crop, overall improvements in water RUE can occur.

Over the long term, reduced soil erosion maintains productivity of the soil. Langdale et al. (1979) showed that loss by water erosion of 150 mm of topsoil on a Southern Piedmont soil led to 40% annual yield reductions in corn. Buntley and Bell (1976) found similar erosional losses in Tennessee reduced corn, soybean, wheat, and tall fescue (*Festuca arundinacea* Schreb.) yields by 42, 50, 28, and 25%, respectively. Multiple cropping can enhance soil protection during critical erosion periods through increased vegetative cover (Siddoway and Barnett, 1976).

Crops are not equal in their conversion efficiencies, defined as production of biomass per unit of water captured. Inclusion of more efficient crops by using species that are best adapted to each growing season can increase overall water RUE. In this way, Belesky et al. (1981) were able to produce more forage biomass without additional water by planting a warm-season annual during the summer and a cool-season annual during the winter into a tall fescue sod, which produces most of its growth in spring and fall. Tall fescue alone produced 7405 kg ha^{-1} over 3 yr at this Southern Piedmont location. When they killed half of the fescue sod in strips and planted sorghum-sudangrass (*Sorghum bicolor* L. Moench × *Sorghum bicolor* L. Moench), total average forage yields increased to as much as 9564 kg ha^{-1}. By planting rye during the fall, total average forage yields increased to 11 115 kg ha^{-1}. Water RUE for the year increased 50% by using a sequential/intercrop system that incorporated species which were better adapted and thus more efficient converters of water and other inputs.

Enhanced Use of Other Resources

When one resource is limiting growth or development, other nonlimiting resources are usually used less efficiently. If the limitation to growth is alleviated, all resources will be used more efficiently until another resource becomes limiting. Since water is often the most limiting resource, any improvement in water RUE due to multiple cropping may also lead to enhanced use of other resources.

Efficiencies by Maximizing Use of Land, Irrigation System, Soil

The different water RUE described above consider what can happen within the boundaries of a field over the entire year when multiple cropping is used. Expanding to the broader perspective reveals an additional source of efficiency. Assuming all crops are utilized, double cropping can decrease the amount of land necessary to produce those crops. Stationary irrigation systems are more effectively utilized, and highly productive land can be used rather than less productive, more erosive land. This means that less area of a watershed or groundwater recharge zone may be affected by cropping practices. Forbes et al. (1984) used a multicrop rotation of cabbage (*Brassica oleracea* L.) in winter, sweet corn in spring, and soybean in summer to maximize the use of old vegetable land in southern Florida. The land was normally idled during summer months because heat and pests made vegetable production unfeasible. They noted that K applied to the cabbage was sufficient for the sweet corn and of benefit to the soybean. Thus, in addition to improvements in land use, the multiple crop system improved capture and conversion of the K which is mobile in these sandy soils.

NITROGEN RESOURCE-USE EFFICIENCY IN MULTIPLE CROPPING

Increased Use of Residual Nitrogen

As with water, one source of N RUE that can be gained through multiple cropping is capture and conversion of that portion of the resource that would be lost during the off-season. Decomposition of residues during the winter may lead to mineralization and nitrification. In climates where rainfall exceeds ET during the off season, nitrates may leave the boundaries of the production unit in runoff and groundwater recharge. For example, in some portions of the Southeast soil temperatures rarely fall below freezing, and mineralization can occur year round. A monoculture of maize will produce biomass for 165 d or less. Thus, more than 200 d remain each year during which decomposition of those residues can proceed. This mineralized N could easily leach from the soil by the high rainfall common in the region. Unfertilized rye grown as a winter cover was able to capture and retain an average of 36 kg of N ha^{-1} in aboveground biomass prior to no-till planting of maize (Ebelhar et al., 1984). In addition to leaching losses during the off-season, mineralized N may be lost through denitrification. Off-season crops may capture at least some of this N before denitrification can occur, and it may lower soil water content sufficiently to lower denitrification rates.

Producing a crop on the N which would otherwise be lost would result in improved N RUE. However, relying upon this N for a second crop is risky; proper water and temperature conditions must be present to allow decomposition and mineralization to occur. The rate of mineralization may be slow enough that N availability limits biomass production. As discussed earlier, this limit would reduce the efficiency of other inputs to crop production.

Huntington et al. (1985) examined recovery of N from winter annual rye and hairy vetch (*Vivia villosa* Roth) by maize. They noted that there was a poor synchronization between time of N release from the cover crop residues and peak maize uptake period. The majority of the N became available only after silking in the maize. While hairy vetch was more effective than rye in supplying N, the recovery could be improved with better synchronization of the legume decomposition with the rapid crop uptake period.

Capture of Fertilizer Nitrogen

An important way in which multiple cropping may improve N RUE is the capture of excess fertilizer N. This occurs in both intercrop and sequential multiple cropping. Capture of what would otherwise be lost in the off-season, in the off-use period, or even during the uptake period, occurs because of different rooting habits, uptake periods, or capture efficiency.

Miscellaneous Improvements in Nitrogen Capture or Conversion Efficiency

Just as off-season crops improve surface residues and soil structure leading to general improvement in water RUE for the main crop, they may also lead to improvement in N RUE. And, N which is captured off-season and retained in residues remains as part of the resource base to supply future crops. The availability of this residue N to succeeding crops will depend upon the decomposition of the residue. This is affected by its C/N ratio and lignin content (Bruulsema and Christie, 1987; Herman et al., 1977) as well as soil environmental conditions. As the retained N is captured and converted to biomass, N RUE improves. Hargrove et al. (1983) found higher N concentrations in winter wheat following soybean than following sorghum in wheat double-crop systems. Using soybean rather than sorghum in the rotation was equivalent to supplying 20 to 34 kg of N ha^{-1} to the wheat.

Systems Including Legumes

A multiple-cropping practice regaining popularity is the use of legumes as the second crop to provide N to the main crop. Hairy vetch grown as a winter cover was capable of supplying 90 to 100 kg of N ha^{-1} to the corn crop which followed in Kentucky and Delaware studies (Ebelhar et al., 1984; Mitchell and Teel, 1977). Any fixed N increases the total resources available to the system just as fertilizer N does. On the other hand, Ladd et al. (1983) found that legumes made only a small contribution to the subsequent crop of wheat. They suggested that the main value of legumes may be in the long-term maintenance of soil organic nitrogen.

When the legume can be harvested, additional biomass is removed, and total N RUE may increase. When the legume is not harvested, the fixed N replaces fertilizer N as an input to the resource base, but, strictly speaking it does not improve total N-use efficiency. There may be economic advantages to producing N using legumes; on the other hand, planting and production costs of the legume can outweigh any economic advantage.

Intercropping systems that include a legume have the potential for transferring some of the fixed N to the associated nonlegume crop (Ismaili and Weaver, 1986; Ta and Faris, 1987). The capture of this fixed N by the nonlegume can increase total N RUE. Hiebsch and McCollum (1987) showed that with intercrops which include a legume/nonlegume mixture, a greater use of land and growing season results than when the two are produced in monocultures. They noted that the legume/nonlegume intercrop system was one of the few systems to show an intercrop production advantage that was not associated with complementary use of time. The increased ATER occurred when the nonlegume crop was grown under low-N conditions. As fertilizer or soil N increased, production of the nonlegume species increased, and the mixture produces no more than the monocultures.

DECREASES IN WATER AND NITROGEN RUE WITH MULTIPLE CROPPING

The foregoing section described how multiple cropping may result in improvements in water and N RUE. Multiple cropping can also result in decreases in those efficiencies.

Water from seasonal and annual rainfall is often limited in comparison with the amount that can be utilized for other purposes. Recharge to groundwater aquifers is necessary to replace withdrawals made for municipal, industrial, and agricultural use. Recharge to streams through runoff and interflow maintains a base flow which supplies downstream users, dilutes contaminants, and supports native vegetation and wildlife. Besides these off-site effects, the leaching of excess or imbalanced salts present in the root zone is diminished from the conversion of these off-season rains to evaporation.

Even when no value is assigned to alternative uses of excess rainfall, production of a second crop during the off-season could result in depletion of soil water by planting time of the main crop. While the rainfall may have been effectively utilized, the resulting depletion increases risks to stand establishment and early season growth. Any yield reduction resulting from it could lower the total water RUE. This would be particularly important in semiarid regions where off-season rains are a primary source of the crop water supply.

Murdock and Wells (1978) noted that in some years use of a small grain silage during winter led to lower yields of corn in the double-crop silage sequence, than corn in a monocrop sequence. They suggested that the actively growing small grain crop removed moisture from the soil in years with dry spring weather, leaving less for the corn. Even so, total silage yields for the double-crop system were 19 to 31% greater than the monocrop system. Ebelhar et al. (1984) noted that the higher-producing cover crops decreased soil moisture content at the time of planting corn. The lower moisture did not affect germination of corn in their study, but they noted that it could be a problem in a dry spring.

Adem and Tisdall (1984) reported that a sequential cropping system which included tomato resulted in a 36% decrease in aggregate stability and 14% decrease in organic carbon as compared with permanent pastures. The destruction of aggregate stability by tillage for the tomato production led to reductions in infiltration.

An intercrop system in which corn was planted into strips of tall fescue that were killed with herbicides was compared with a sequential cropping system of winter rye followed by corn (Wilkinson et al., 1987). The intercrop system resulted in a greater reduction in corn plus forage biomass than when the crop was planted as a sequential crop. Competition for water and N by the grass and corn led to 10 to 48% lower yields than when the competition was eliminated. Presumably water and N RUE were reduced as well.

SUMMARY

Multiple-cropping systems are those that result in an average of more than one crop per year on a unit of land. Multiple cropping may involve sequential cropping, intercropping, or combinations of the two. Because multiple-cropping systems extend beyond the planting to harvest limits of one crop, alternative definitions are necessary in order to evaluate their water and N efficiencies. Resource-use efficiency is a systems concept that can be useful in evaluating production efficiencies of the multiple-crop farm system. For water and N, all resources within the boundaries and all inputs to the system are the basis for efficiency computations. Biomass or harvested production per available resources are examined, although economic value of that produced ultimately determines resource-use efficiency. Values can be assigned to outputs other than crop products for a more comprehensive RUE examination.

Multiple-cropping systems are most effective in improving water and N RUE when those systems capture and convert water and N that would otherwise be lost during noncrop periods of the monocrop culture. By increasing total production, multiple cropping can reduce the land requirements needed for a given quantity of products. While water and N RUE may be improved by multiple cropping, they also may be decreased.

Improvement of N and water RUE through multiple-cropping systems is by itself insufficient justification for establishing multiple-cropping systems. That justification must come from a socioeconomic analysis of the systems. When water and N are underutilized or lost by the practice of one crop per year, the added RUE afforded by multiple cropping suggests that they should be considered in the analysis.

REFERENCES

Adem, H.H., and J.M. Tisdall. 1984. Management of tillage and crop residues for double-cropping in fragile soils of Southeastern Australia. Soil Till. Res. 4:577–589.

Beale, O.W., G.B. Nutt, and T.C. Peele. 1955. The effects of mulch tillage on runoff, erosion, soil properties and crop yield. Soil Sci. Soc. Am. Proc. 19:244–247.

Belesky, D.P., S.R. Wilkinson, R.N. Dawson, and J.E. Elsner. 1981. Forage production of a tall fescue sod intercropped with sorghum × sudangrass and rye. Agron. J. 73:657–660.

Bruulsema, T.W., and D.R. Christie. 1987. Nitrogen contribution to succeeding maize from alfalfa and red clover. Agron. J. 79:96–100.

Buntley, G.J., and F.F. Bell. 1976. Yield estimates for the major crops grown on soils of West Tennessee. Tennessee Agric. Exp. Stn. Bull. 561.

Doorenbos, J., and A.H. Kassam. 1979. Yield response to water. FAO Irrig. Drain. Paper 33. FAO of the United Nations, Rome.

Ebelhar, S.A., W.W. Frye, and R.L. Blevins. 1984. Nitrogen for legume cover crops for no tillage corn. Agron. J. 76:51–55.

Forbes, R.B., J.B. Sartain, and N.R. Usherwood. 1984. Optimum K fertilization schedule for maximizing yields of cabbage, sweet corn and soybeans grown in a multicropping sequence. Proc. Soil Crop Sci. Soc. Fla. 43:64–68.

Francis, C.A. (ed.) 1986. Multiple cropping systems. Macmillan Publ. Co., New York.

Frank, A.B., R.E. Barker, and J.D. Berdahl. 1987. Water use efficiency of grasses grown under controlled and field conditions. Agron. J. 79:541–544.

Hammond, L.C., R.S. Mansell, W.K. Robertson, J.T. Johnson, and H.M. Selim. 1981. Irrigation efficiency and controlled root-zone wetting in deep sands. Florida Water Resour. Res. Ctr. Publ. 52.

Hanks, R.J., H.R. Gardner, and R.L. Florian. 1969. Plant growth-evapotranspiration relations for several crops in Central Great Plains. Agron. J. 61:30–34.

Hargrove, W.L., J.T. Touchton, and J.W. Johnson. 1983. Previous crop influence on fertilizer nitrogen requirements for double-cropped wheat. Agron. J. 75:855–859.

Herman, W.A., W.B. McGill, and J.F. Domaar. 1977. Effects of initial chemical composition on decomposition of roots of three grass species. Can. J. Soil Sci. 57:205–215.

Hiebsch, C.K., and R.E. McCollum. 1987. Area × time equivalency ratio: A method for evaluating the productivity of intercrops. Agron. J. 79:15–22.

Hildebrand, P.E. 1976. Multiple cropping systems are dollars and "sense" agronomy. p. 347–371. *In* R.I. Papendick et al. (ed.) Multiple cropping. Spec. Publ. 27. ASA, Madison, WI.

Hillel, D., and Y. Guron. 1973. Relation between evapotranspiration rate and maize yield. Water Resour. Res. 9:743–749.

Huntington, T.G., J.H. Grove, and W.W. Frye. 1985. Release and recovery of nitrogen from winter annual cover crops in no-till corn production. Commun. Soil Sci. Plant Anal. 16:193-211.

IRRI. 1974. Multiple cropping. p. 15-24. *In* Annual report for 1973. IRRI, Los Banos, Philippines.

Ismaili, M., and R.W. Weaver. 1986. Competition between Sirato and kleingrass for ^{15}N labeled mineralized nitrogen. Plant Soil 96:327-335.

Jones, J.N., Jr., J.E. Moody, and J.H. Lillard. 1969. Effects of tillage, no-tillage, and mulch on soil water and plant growth. Agron. J. 61:719-721.

Ladd, J.N., M. Amato, R.B. Jackson, and J.H. Butler. 1983. Utilization by wheat of nitrogen from legume residues decomposing in the field. Soil Biol. Biochem. 15:231-238.

Langdale, G.W., J.E. Box, Jr., R.A. Leonard, A.P. Barnett, and W.G. Fleming. 1979. Corn yield reduction on eroded Southern Piedmont soils. J. Soil Water Conserv. 34:226-228.

Michell, W.H., and M.R. Teel. 1977. Winter annual cover crops for no-tillage corn production. Agron. J. 66:569-573.

Moody, J.E., J.N. Jones, Jr., and J.H. Lillard. 1963. Influence of straw mulch on soil moisture, soil temperature and growth of corn. Soil Sci. Soc. Am. Proc. 27:700-703.

Murdock, L.W., and K.L. Wells. 1978. Yields, nutrient removal and nutrient concentrations of double cropped corn and small grain silage. Agron. J. 70:573-576.

Musick, J.T., and D.A. Dusek. 1980. Irrigated corn yield response to water. Trans. ASAE 23:92-98, 103.

Papendick, R.I., P.A. Sanders, and G.B. Triplett (ed.) 1976. Multiple cropping. Spec. Publ. 27. ASA, Madison, WI.

Siddoway, F.H., and A.P. Barnett. 1976. Water and wind erosion aspects of multiple cropping. p. 317-335. *In* R.I. Papendick et al. (ed.) Multiple cropping. Spec. Publ. 27. ASA, Madison, WI.

Stegman, E.C., J.T. Musick, and J.I. Stewart. 1980. Irrigation water management. p. 763-816. *In* M.E. Jensen (ed.) Design and operation of farm irrigation systems. ASAE Monogr. 3. ASAE, St. Joseph, MI.

Ta, T.C., and M.A. Ferris. 1987. Effects of alfalfa proportions and clipping frequencies on timothy-alfalfa mixtures. II. Nitrogen fixation and transfer. Agron. J. 79:820-824.

Trenbath, B.R. 1974. Biomass productivity of mixtures. Adv. Agron. 26:177-210.

----. 1976. Interactions among the components of intercrops. Plant interactions in mixed crop communities. p. 129-169. *In* R.I. Papendick et al. (ed.) Multiple cropping. Spec. Publ. 27. ASA, Madison, WI.

----. 1986. Resource use by intercrops. p. 57-81. *In* C.A. Francis (ed.) Multiple cropping systems. Macmillan Publ. Co., New York.

Wilkinson, S.R., O.J. Devine, D.P. Belesky, J.W. Dobson, Jr., and R.N. Dawson. 1987. No-tillage intercropped corn production in tall fescue sod as affected by sod-control and nitrogen fertilization. Agron. J. 79:685-690.

Willey, R.W. 1985. Evaluation and presentation of intercropping advantages. Exp. Agric. 21:119-133.

Zweifel, T.R., J.W. Maranville, W.M. Ross, and R.B. Clark. 1987. Nitrogen fertility and irrigation influence on grain sorghum nitrogen efficiency. Agron. J. 79:419-422.

3 Crop Rotation and Its Impact on Efficiency of Water and Nitrogen Use[1]

Francis J. Pierce and Charles W. Rice

Michigan State University
East Lansing, Michigan

Crop rotations and their benefits to agriculture have long been known. Like many other topics in agriculture, the role of crop rotations is subject to many truisms seemingly substantiated by a great many observations over a long time. The scientific and popular literature provide a number of explanations for the general observation that crops grown in rotation often yield more than when grown in monoculture (Arnon, 1972; Curl, 1963; Mannering and Griffin, 1981). Certain crops in rotation, however, can be detrimental to succeeding crops through increased pest problems, allelopathy, or high-consumptive water requirements. Although the use of crop rotations declined with the introduction of fertilizer and pest and disease control chemicals, problems do exist with monoculture (USDA, 1973). In addition, there is renewed interest in rotations in the 1980s associated with the following developments.

1. In today's era of decreased commodity prices, renewed interests in rotations has evolved as a means of increasing farm profitability, in part through diversification.
2. Yields of crops grown using conservation tillage systems, particularly no-tillage, may improve when crops are grown in rotation rather than in monoculture (Mannering and Griffin, 1981; Triplett, 1986). Therefore, increased adoption of conservation tillage systems should increase interest in crop rotations.
3. Concerns over agriculture's impact on environmental quality, particularly surface and groundwater quality, continue to mount. Systems that reduce chemical inputs in agricultural production (low input or "sustainable" agriculture) are of growing interest to agronomists, soil scientists, and ecologists. Rotations play an important role in this area as discussed by Schepers (1988) and Radke et al. (1988) in this publication.

---

[1] Contribution of the Michigan State Univ. Agric. Exp. Stn., Paper no. 12542.

Cropping Strategies for Efficient Use of Water and Nitrogen, Special Publication no. 51.

In the future, the need will be for more efficient utilization of our nation's soil, water, and economic resources. The extent to which crop rotations might contribute to that goal is uncertain. The objectives of this chapter are to discuss the role of crop rotations in agricultural production and the ways in which crops grown in rotation impact water and N-use efficiencies in agricultural systems.

TERMINOLOGY

In many cases, the phrases *cropping systems* and *crop sequence* are used interchangeably with *crop rotation.* Baldock et al. (1981) adopted the terminology of Yates (1954) who defines a cropping sequence as a succession of crops grown on a specific land unit and a crop rotation (or simply *rotation*) as a specific type of crop sequence in which the succession of crops is repeated. This discussion will consider all cropping sequences under the term rotation although cover crops, intercrops, and multiple crops are discussed in more detail elsewhere in this publication. Sequences of crop production where *fallow* is in annual sequence with a crop will also be considered under the term rotation.

The difference in yield associated with the rotation of crops is often termed the *rotation effect.* As illustrated in Fig. 3-1, corn (*Zea mays* L.) following another crop in rotation may result in significantly higher corn grain yields than corn following corn. It is common to attribute some or all of the effect of rotation to one or more factors. For example, the portion

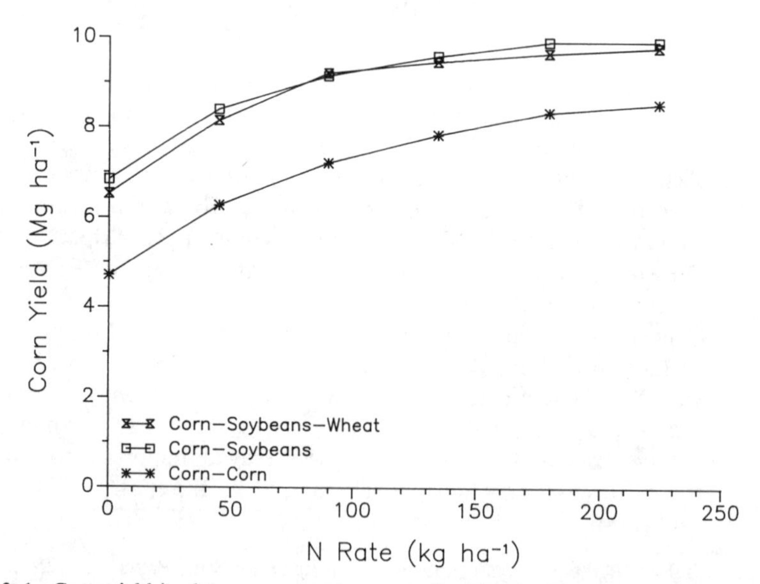

Fig. 3-1. Corn yield in three crop rotations as influenced by N rate on a Webster clay loam (fine-loamy, mixed, mesic Typic Haplaquoll) at Waseca, MN averaged for 1975 to 1986 (Randall, 1987).

of the yield increase attributable to the N contribution from the previous crop could be termed the *N rotation effect*. Baldock et al. (1981) and Hesterman et al. (1987), however, reserved the term *rotation effect* to include only that portion of the yield difference not attributable to some known factor.

In their view, in terms of yield, the total rotation effect is the yield differential at zero N applied, the rotation effect is the yield differential at the highest N-fertilization rate, and the N-rotation effect is the total minus the rotation effect. In Fig. 3–1, in the corn-soybean [*Glycine max* (L.) Merr.] rotation, the total rotation effect is 2.1 Mg ha^{-1}, the rotation effect is 1.4 Mg ha^{-1}, and the N-rotation effect is 0.7 Mg ha^{-1}. When fertilizer additions result in equal yields for rotation and monoculture, the rotation effect is zero. Where rotations result in decreased yields, the rotation effect is negative.

ROTATIONS AND NITROGEN-USE EFFICIENCY

Recent papers discuss N-use efficiency (NUE) in crop production (Hauck, 1984), specifically with regards to crop rotations (Kurtz et al., 1984) and cropping systems (Bock, 1984). Research involving crop rotations have typically addressed the fertility aspects of rotations, especially those in the first half of this century, and often involve N-rate studies and their combined impact on crop production. Because crops in rotation can provide some or all of the N requirement of the succeeding crop, fertilizer-use efficiencies can, in some cases, be reduced relative to monoculture. However, few studies are able to quantify the N-use efficiencies of crops in rotation because N is available from a variety of N pools in soils. This discussion focuses on the impact of crop rotations on NUE in the broad sense and the impact of rotations on the use efficiencies of the various available N pools in soil.

The discussion of the impact of crop rotations on NUE is affected by a number of factors, not the least of which can be found in the definition of NUE. Nitrogen-use efficiency is often defined as the ratio of crop yield or crop N uptake from a specific source to the total soil input of that source. Hesterman et al. (1987) used that concept of NUE to define the use efficiency of legume N and fertilizer N. Bock (1984) defined NUE in terms of yield efficiency (yield/N applied = Y/N_f), recovery efficiency (N recovered/N applied = NR_f/N_f), and physiological efficiency (yield/N recovered = Y/NR_f), where yield efficiency is the product of the N recovery and physiological efficiencies.

$$\frac{Y}{N_f} = \frac{NR_f}{N_f} \times \frac{Y}{NR_f} \qquad [1]$$

Although well defined and evaluable where N is applied, these definitions are lacking in the case of rotations, since crop uptake of N comes from diverse sources. When $N_f = 0$, Eq. [1] is undefined. In addition, crop yields may not be affected by applied N. Figure 3–2 illustrates six possible fertilizer-response curves comparing crops grown in rotation vs. monoculture. Figure

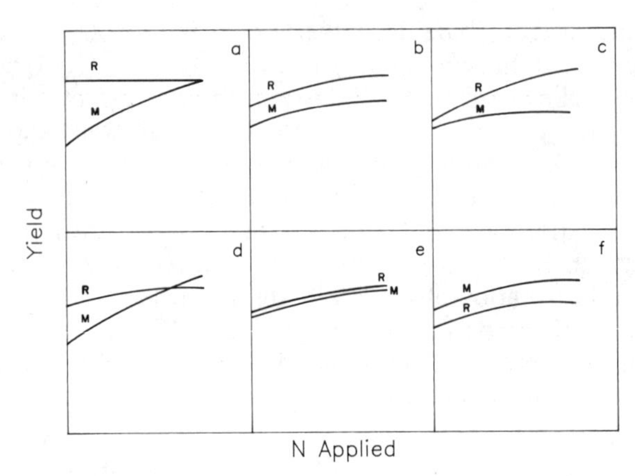

Fig. 3–2. Six types of crop responses to applied N comparing rotation (R) to monoculture (M).

3–2a represents a situation where yields in rotation do not respond to N fertilization. Yields in monoculture are below those in rotation but equal those in rotation at high rates of N applied. Blackmer (1987) reports this response where corn yields following alfalfa showed no response to N but yields were equal to yields under continuous corn at high N rates. From Eq. [1], Y is a constant in rotation and yield efficiency will decrease with increased N applied. Figure 3–2a illustrates three points: (i) fertilizer-use efficiency in rotation is undefined unless tracer N is used to determine fertilizer-N uptake, (ii) the so-called rotation effect is often attributed to an *N effect*, and (iii) N-use efficiency is undefined at zero N applied for either rotation or monoculture when NUE is defined in terms of N applied.

A major point illustrated by each curve in Fig. 3–2 is that NUE is somewhat undefined. Therefore, a more inclusive definition of NUE is needed to encompass all sources of N available to plants, although its application may be difficult.

Meisinger (1984) used mass balance principles to describe plant-available N for any soil-crop system. The following definition of NUE utilizes the mass balance approach:

$$\text{NUE} = \frac{\text{Plant N uptake}}{\Sigma \text{ N available pools}} = \frac{N_p}{N_a} . \qquad [2]$$

It would be most desirable to express plant uptake, N_p, in terms of total biomass N. Roots, residues, and root slough materials are difficult to measure and most often ignored. Aboveground biomass N, therefore, would be the most plausible estimate of N_p, although harvestable yield appears to be most common.

The N sources and pools available for uptake are given in Table 3–1. Using the symbols for each N pool, Eq. [2] takes the form:

Table 3-1. Nitrogen sources and pools available for plant uptake.

Nitrogen source and pools†	Symbol
Fertilizer	N_f
Crop residue	N_c
Inorganic N	N_i
Mineralizable N	N_m
a. Microbial biomass	N_b
b. Soil organic matter	N_s
c. Other (fauna, etc.)	N_z
Fixed N	N_x
a. Symbiotic	N_{xs}
b. Nonsymbiotic	N_{xn}
Depositional (atmospheric, irrigation, runon, etc.)	N_d
Nitrogen outputs	
Plant uptake	N_p
Nitrogen lost by erosion	N_{er}
Nitrogen lost by leaching	N_l
Gaseous N losses	N_g
Nitrogen reacted	
Nitrogen immobilized	N_{im}
Nitrogen chemically fixed	N_{cf}

† $N_m = \{N_b + N_s + N_z\}$ and $N_x = \{N_{xs} + N_{xn}\}$.

$$NUE = \frac{N_p}{[N_c + N_i + N_m + N_x + N_d]}. \quad [3]$$

When expanded to include subpools, NUE takes the form:

$$NUE = \frac{N_p}{[N_f + N_c + N_i + N_b + N_s + N_z + N_{xs} + N_{xn} + N_d]}. \quad [4]$$

The use efficiency of each pool is as follows:

$$NUE_{pool} = \frac{Nu_{pool}}{N_{pool}}, \quad [6]$$

where Nu indicates uptake of N from any N_{pool}. Nitrogen-use efficiency can be described in terms of the efficiency of use of each N pool as follows:

$$NUE = \frac{\Sigma(NUE_{pool})(Nu_{pool})}{N_f + N_c + N_i + N_b + N_s + N_z + N_{xs} + N_{xn} + N_d}. \quad [7]$$

Nitrogen fertilizer-use efficiency (NUE_f) is a common parameter discussed in reference to NUE. Under this terminology, NUE_f would be expressed as

$$NUE_f = \frac{Nu_f}{N_f} \quad [8]$$

d is equivalent to Bock's (1984) recovery efficiency.

The amount of N available to a plant (N_a) according to the whole crop N balance is the denominator of Eq. [4]. These N_a pools are also subject to processes that lead to losses from the system or immobilization of once available N described in Table 3-1 as N outputs and N reacted. It is clear that rotations affect the processes that remove N from available pools (discussed later). Nitrogen-use efficiency overall, or for any subset NUE_{pool}, is affected by crop demand, soil parameters, and a variety of N losses. Changes in the efficiency of use from any N_{pool} due to rotation can occur when either the rotation alters the size of the N_{pool} when compared to monoculture or the rotation stimulates or inhibits the uptake from a given N_{pool}.

In practical terms, if crop rotation increases the pool size of any N_{pool}, NUE_{pool} will increase only if the plant utilizes that pool proportionally greater than that in monoculture.

EFFECTS OF ROTATIONS ON NITROGEN UTILIZATION

Studies by Abshahi et al. (1984) and Hesterman et al. (1987) serve as examples of efforts to quantify a direct effect of rotation, that of the N contribution of the previous crop to a succeeding crop. Neither study represents the true sense of the term *crop rotation* defined by Yates (1954) but provide excellent examples of the principles involved.

Abshahi et al. (1984) quantified wheat utilization of N from residual N fertilizer and sugarbeet (*Beta vulgaris* L.) tops. Their data are summarized in the mass balance format in Table 3-2. They applied four rates of fertilizer N (0, 62, 124, and 186 kg of N ha^{-1}) with and without sugarbeet residue to wheat (*Triticum aestivum* L.) planted after sugarbeets. They determined the sugarbeet residue (N_c) to contain 117 kg of N ha^{-1}. Since most N_{pools} could not be determined, we estimated the N_{soil} pool as the maximum soil N contribution to plant uptake determined across all treatments (118 kg of N ha^{-1}). The NUE_f ranged from 82 to 59%. Nitrogen-use efficiency ranged from 80 to 64% and was higher than NUE_f with the exception of the treatment where 62 kg of N ha^{-1} with residue applied had NUE_f of 82%. Both NUE and NUE_f decreased with increasing fertilizer applied. The addition of sugarbeet residues decrease NUE assuming all residue N was available. Fertilizer economic yields were achieved in the treatments receiving the 62 and 124 kg of N ha^{-1} fertilizer rates with sugarbeet residues. The most N-use efficient and economical treatment, therefore, was the residue treatment with 62 kg of N ha^{-1}.

Hesterman et al. (1987) investigated the short-term rotation effects of legumes on corn for two soils in Minnesota. Data available from this publication were not sufficient to construct the N balance table. However, the authors report N-use efficiencies for fertilizer and legume N. NUE_f for $(NH_4)_2\ SO_4$ fertilizer averaged 47 and 54% at the Becker and Rosemount, MN locations, respectively. Legume N-use efficiencies by corn ranged from 27 to 66% and

Table 3–2. Nitrogen mass balance estimated from the rotation N study of Abshahi et al. (1984).

	Fertilizer rate, kg ha^{-1}							
	0	0	62	62	124	124	186	186
	Sugarbeet residue							
	NR†	R	NR	R	NR	R	NR	R
	Inputs, kg ha^{-1}							
N_{pool}								
N_f	0	0	66	21	24	124	186	186
N_c	0	117	0	117	0	117	0	117
N_{soil}‡	118	118	118	118	118	118	118	118
	Outputs, kg ha^{-1}							
N_p	94	141	134	192	175	226	215	243
N_{fu}	0	0	43	51	76	76	109	102
N_{cu}	0	32	0	33	0	33	0	33
N_{soilu}§	94	109	91	101	92	118	106	108
	Nitrogen-use efficiencies, %							
NUE upper limit	80	60	74	65	72	63	71	58
NUE_f	--	--	69	82	61	61	59	55
NUE_c	--	27	--	28	--	28	--	28
	Grain yields, Mg ha^{-1}							
	4.2	5.8	4.9	6.6¶	6.5	6.9¶	6.7	6.8

† NR = No sugarbeet residue, R = sugarbeet residue added.
‡ N_{soil} is the estimated soil contribution that was the maximum calculated across all N rates. $N_{soil} = f(N_c, N_i, N_b, N_s, N_{nx}, \text{and } N_d)$.
§ N_{soilu} = estimate of N uptake by plant from soil.
¶ Fertilizer N for optimum yield.

39 to 70% for Becker and Rosemount, respectively, depending on alfalfa (*Medicago sativa* L.) variety and frequency of cutting. Legume N-use efficiencies from soybean by whole plant corn averaged 18 and 39% and 19 and 50% for nodulating and nonnodulating varieties at each location, respectively. The legume N-use efficiencies reported are less than NUE_c given in Table 3–1 since NUE_c is calculated using the available N from that N pool which would be less than the legume N returned to the soil. The NUE_c's reported by Hesterman et al. (1986) for legumes therefore, do not represent the maximum according to Eq. [6]. The NUE_c's for nodulating soybean are lower than for alfalfa or nonnodulating soybean. These results illustrate the potential impact of genetic variation in crop management strategies on N_{pool} and NUE.

Another effect of rotation on N-use efficiency is found in wheat-fallow rotation of the Great Plains region of the USA. Water conservation is a primary objective of the fallow system. However, the fallow period is also critically important to NO_3-N accumulation, the rate of which significantly affects the interpretation of the residual soil NO_3-N test (Lamb et al., 1985). At the Alliance, NE site, NO_3-N accumulation during fallow from 1970 to

1983 ranged from 12 to 135 kg of N ha^{-1} when no fertilizer was applied and 22 to 150 kg of N ha^{-1} when 45 kg of N ha^{-1} was applied as fertilizer to wheat (Lamb et al., 1985). Nitrate accumulation was higher when fertilizer was applied and differed with soil tillage management. Nitrogen mass balance could not be determined from this data. However, the large pools of inorganic nitrogen in the profiles would certainly impact use efficiencies of the N pools in this rotation.

EFFECTS OF CROP ROTATION ON NITROGEN POOLS AND SOURCES

Effects of rotations on N-use efficiency are difficult to establish since N entering the soil system is subject to a variety of chemical reactions, biological transformations, and loss mechanisms. Impacts of crop rotations on NUE may be indirect and are primarily involved with changes in the various N pools and sources in both quantity and plant availability. These effects are generally the result in long-term (>1 yr) impacts on N utilization.

The following discussion examines the impact of rotations on N sources and pools (Table 3–1) giving specific examples to illustrate each impact.

N_c

The type of residue from the previous crop will influence the amount and rate of N mineralization from crop residue. Corn following alfalfa has been the most frequently studied; however, few studies have quantified the actual amount of N_c released and uptake by a succeeding crop. Hesterman et al. (1987), using ^{15}N-labeled alfalfa residues, determined the amount of N_c recovered by corn in the succeeding year; however, they did not estimate the total quantity of legume N mineralized during the first crop year (N_c in Table 3–1) or the residual N of the incorporated legume residues in succeeding years. Ladd et al. (1981a, b) has provided the most complete study on the fate of legume N. Only 5 to 10% of medic residue N was utilized by a succeeding wheat crop (Ladd et al., 1981b). They concluded that the primary benefit of legumes is a long-term effect of maintaining soil organic nitrogen. Other studies have used N-rate response curves to derive the "rotation N benefit" (Levin et al., 1987; Triplett et al., 1969). Levin et al. (1987) estimated that the soil and residue together supplied 126 kg of N ha^{-1}, however they were unable to separate soil and residue as N sources, a situation typical of most studies.

Sorghum [*Sorghum bicolor* (L.) Moench] residues, because of a low N content, not only release N more slowly than alfalfa but may initially immobilize soil inorganic nitrogen (N_i) and fertilizer N (N_f) during decomposition. Immobilization of soil and fertilizer N by sorghum residue can result in N-deficient doublecropped wheat (Sanford et al., 1973; Hargrove et al., 1983). Similarly, a doublecrop soybean yield reduction due to standing wheat residue was overcome by a small addition of N fertilizer (Hairston et al.,

1987). Wagger et al. (1985) examined the pattern of wheat and sorghum residue mineralization using ^{15}N-labeled residue in monoculture. Sorghum residue initially mineralized N at a slower rate, however, the greatest N release occurred during the period of greatest N demand by a subsequent sorghum crop, i.e., synchrony. Their results indicate that a crop immediately following sorghum would require additional N because of immobilization or low initial mineralization. For wheat residue, the total amount of N released and subsequently recovered by a second wheat crop was similar to that observed in sorghum monoculture; however, the wheat-residue N mineralization rate was initially faster than that of sorghum residue N.

N_i

Inorganic nitrogen in the soil profile will be affected by the frequency of crops receiving large N-fertilizer inputs. Olsen et al. (1970) found a direct relationship between the total NO_3 in the soil profile and the frequency of corn and N fertilizer applied in the rotation. Thus, continuous corn would leave more soil inorganic nitrogen in the profile than rotations. In wheat-fallow rotations, one of the benefits of fallow is the accumulation of NO_3 in the soil profile that is available for utilization by the following wheat crop (Lamb et al., 1985). The increased quantity of soil NO_3 can also have negative impacts due to increased leaching losses, groundwater contamination, and reduced symbiotic N_2 fixation by succeeding legume crops.

N_{xs}

The rotation effect on symbiotic N_2 fixation is important because of an expected increased future reliance on and benefit from biological fixed N in cereal-legume rotations and many low-input agroecosystems. Rotations can affect symbiotic N_2 fixation by two mechanisms. Rotation may effect the survival of the microbial symbiont, i.e., rhizobia. In a cotton (*Gossypium hirsutum* L.)-corn-soybean rotation, populations of *Bradyrhizobium japonicum* were able to survive in the intervening years when the host was absent at sufficient populations to potentially reinfect soybean unless the soil pH was <4.6 (Hiltbold et al., 1985). Unfortunately, the survival of effective strains and their reinfectivity were not quantified. The second mechanism of crop rotations effects on N_2 fixation involves the control of residual inorganic nitrogen from the previous crop. A vast amount of literature exists concerning inorganic nitrogen effects on N_2 fixation. But, the role of rotations has seldom been examined in detail. Bezdicek et al. (1974) did examine rotation effects on N_{xs} in detail and reported a decrease in nodule mass and N_2-fixation rate in soybean with the addition of inorganic nitrogen fertilizer. When N fertilizer was added to a preceding rye (*Secale cereale* L.) cover crop, the rye removed sufficient inorganic nitrogen resulting in increased N_2 fixation and soybean grain yield. Legumes can also act as a "catch crop" for soil inorganic nitrogen thus conserving N in the soil and reducing leaching and denitrification. If the goal is to increase biologically fixed N input into the soil, it is important to consider residual inorganic nitrogen at the time of legume planting to maximize N_2 fixation.

N_{ns}

Nonsymbiotic N_2 fixation has not been examined to a great extent in soil microbial ecology, but is usually considered small in proportion to the other N sources, particularly in N-fertilized systems. Lamb et al. (1987) measured a twofold increase in nonsymbiotic N_2 fixation in no-till soil compared to plowed soil for a total of <1 kg ha^{-1} yr^{-1}. They found a strong relationship between N_2 fixation and soil moisture. It might be reasoned that organic carbon would also increase the potential for nonsymbiotic fixation as this increases microbial activity. Crop rotations that increase organic carbon and soil moisture at the surface may result in increase potential for nonsymbiotic N_2 fixation.

N_b

Microbial biomass comprises an active N pool in soil that relates to soil productivity. The net turnover of this pool can potentially supply N for plant uptake. Only recently has microbial biomass N been considered as a potential for plant N; thus little literature is published on the rotation effects on microbial biomass. Powlson et al. (1987) reported that the addition of wheat straw residue compared to straw burning increased microbial biomass C and N up to 45 and 50%, respectively. They considered the measure of microbial biomass as an early indicator of long-term changes. An "organic" farm of winter wheat-spring pea (*Pisum* spp.)-winter pea (green manure) rotation had significantly greater microbial biomass C and N than a conventional farm of winter wheat-spring pea rotation (Bolton et al., 1985).

An indirect indicator of microbial biomass and its activity is a measure of selected enzyme activities. Higher enzyme activity suggests a more active soil microflora. Bolton et al. (1985) reported increased enzyme activities of urease, phosphatase, and dehydrogenase in the "organic" wheat-spring pea-winter pea rotation compared to the wheat-spring pea rotation previously described. Dick (1984) measured significantly higher enzyme activities of phosphatases, arylsulfatase, amidase, and urease in a corn-oat-alfalfa rotation compared to corn-corn and corn-soybean rotations. Thus, it appears that both greater quantities and turnover of microbial N are associated with some rotations. The potential for exploiting this N pool for more plant N use needs to be examined, particularly in rotations.

N_s

The organic soil nitrogen pool has been the most studied of the N pools in the context of rotations. In general, rotations increase soil organic nitrogen or decrease the rate of soil organic nitrogen loss. In 24-yr wheat-wheat and wheat-fallow rotations, total N contents of 1.1 g of N kg^{-1} and 0.98 g of

N kg^{-1} were measured, respectively (Unger, 1968). Although small, these differences were significant. Juma and McGill (1986) analyzed several long-term rotations. They concluded that continuous cropping resulted in greater turnover rates and higher calculated steady-state levels of soil organic nitrogen than rotations containing fallow. Cultivating and cropping soil generally deplete total soil N, but some rotations will deplete soil organic nitrogen at a greater rate (Burutolu, 1985; Zielke and Christenson, 1986). In order of N conservation, C-C-C-SB = C-C-C-B-SB > O-A-B-SB = C-SB > O-B-SB = B-SB [C = corn, SB = sugarbeet, O = oat (*Avena sativa* L.), A = alfalfa, B = dry bean (*Phaseolus vulgaris* L.)] (Zielke and Christenson, 1986).

A further attempt to refine characterization of total soil organic nitrogen is fractionation by acid hydrolysis (Burutolu, 1985; Stevenson, 1982). Stevenson (1982) summarized several reports on the rotation effect on soil organic nitrogen fractions. In general, amino acid N and amino sugar N increased from C-O < C-C-C < C-O-Cl (Cl = clover [*Trifolium* spp.]) (Stevenson, 1982). Amino acid N is considered to be one of the primary N forms involved in rapid turnover and availability (Ladd and Paul, 1973). Rotations may also affect the chemical distribution among the amino acids (Stevenson, 1982; Young and Mortenson, 1958; Khan, 1971). However, chemical fractionation of soil organic nitrogen has little value in predicting plant-available N (Stevenson, 1982).

Another attempt to refine characterization of the organic nitrogen pool is to determine mineralizable N (N_m). The long-term effect of wheat-fallow rotations can result in less N_m than wheat-fallow rotations; thus, N fertilizer was reportedly necessary to replenish the soil N pool (Janzen, 1987). Different rotations will also affect mineralizable N (Zielke and Christenson, 1986; Bolton et al., 1985). Ladd et al. (1981b) reported N mineralization of the soil organic matter and subsequent uptake by wheat comprised 1 to 2% of the total soil N pool, of approximately 1000 kg of N ha^{-1}. In short-term studies of continuous no-tillage and moldboard plowing, N mineralization and grain yields at low N-fertilizer amounts are often less with no-tillage but comparison of long-term corn tillage systems indicate similar net N mineralization and grain yields due a greater decrease in soil organic nitrogen in moldboard plowed soils (Rice et al., 1986). This result emphasizes the importance of long-term studies to characterize soil N cycling in rotations.

Interactions with Tillage

The dramatic increase in the use of conservation tillage practices have brought with it new interest in the effect of crop rotations on yield, soil physical and chemical properties, NUE, water-use efficiency (WUE), and efficient water use (EWU). Crop residue management at or near the soil surface, along with changes in the type and amount of soil disturbance through tillage, affects many processes controlling agricultural productivity and regulating NUE and WUE. While the role of crop residue management in relation to water and N use are discussed elsewhere in this book (Power, 1988;

Table 3–3. Soybean yields as affected by tillage and crop rotation systems at Crossville, AL for 1983 and 1984 (Thurlow et al., 1985).

	Tillage					
	No-till		Conventional		Strip-tillage	
Rotation	1983	1984	1983	1984	1983	1984
	Mg ha^{-1}					
Continuous soybean	2.2	2.4	0.7	1.8	1.6	2.0
Soybean-corn	2.4	3.0	1.7	1.9	2.2	3.1
Wheat-soybean-corn	1.8	2.4	1.9	2.0	1.9	2.8

Unger, 1988), certain features of the relationship between tillage and rotations are discussed here.

Long-term rotation-tillage studies have shown that crop yields in rotation are generally improved over crops grown continuously. Yield differences between monoculture and rotation corn are higher under conservation-tillage systems than conventional systems depending on soil type (Dick et al., 1986a b; Dick and Van Doren, 1985; Doster et al., 1983; Mannering and Griffin, 1981; Van Doren et al., 1976).

Thurlow et al. (1985) reported yields of soybean that were significantly higher in rotation with corn than when following soybean or wheat (Table 3–3). In conventional-tillage systems, however, soybean yields following wheat were greater than when following corn or soybean. Soybean yields in conservation-tillage systems were higher than in conventional systems except when soybean followed wheat. Data were not included in the study to construct the N mass balance to determine the impact of rotation on NUE and NUE_{pools} in soybean. Because soybean is an effective scavenger of N (Kurtz et al., 1984), the impact on NUE could be significant.

Tillage affects the placement of crop residues, which can affect the N contribution to the succeeding crop. Varco et al. (1985) reported a decrease in the percentage N recovery by corn when hairy vetch (*Vicia villosa* L.) or rye was incorporated by plowing than when left on the soil surface in no-tillage. Nitrogen-use efficiency was greater with no-till corn than with conventional tillage because of greater moisture availability.

There is evidence that long-term rotation effects differ between tillage systems. Dick et al. (1986a, b) reported changes in soil properties after 17 yr of a tillage rotation study. Changes in organic nitrogen as affected by tillage and rotation for the upper 0.2 m of two soils in Ohio are plotted in Fig. 3–3. Patterns of N accumulation (or N loss) were similar for both the Wooster soil (Fine-loamy, mixed, mesic Typic Fragiudalf) and the Hoytville soil (fine, illitic, mesic Mollic Ochraqualf). Organic nitrogen in the surface 75 mm decreased in the order corn-oats-meadow (COM) > corn-corn (CC) > corn-soybean (CS) for both tillage systems, with concentrations in the no-till treatments greater than conventional tillage. There were no significant differences in organic nitrogen below 75 mm in the no-tillage treatments. However, the CS rotation in the conventional tillage treatments had significantly lower organic nitrogen contents than the CC or COM rotation

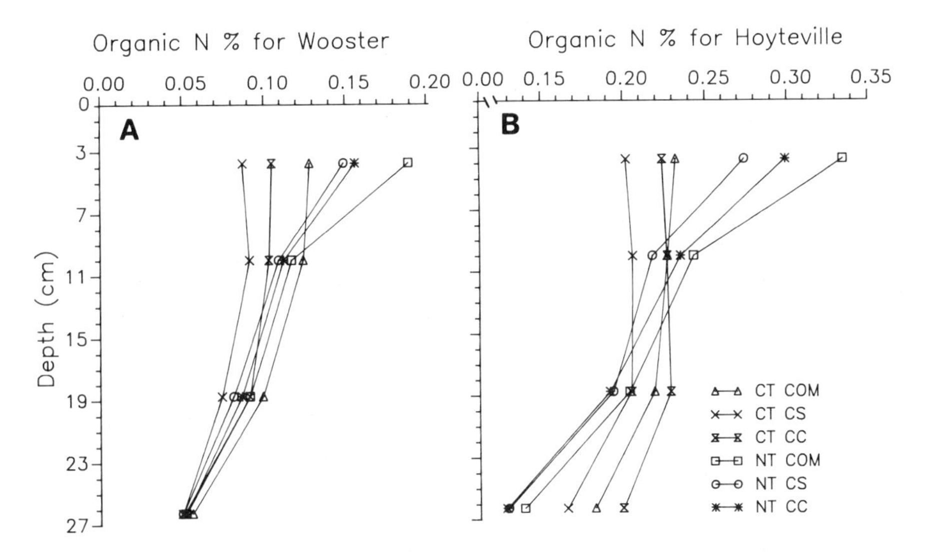

Fig. 3-3. Profile distribution of organic nitrogen after 17 yr of a tillage-crop rotation study on (A) a Wooster soil and (B) a Hoytville soil in Ohio (Dick et al., 1986a, b). Conservation tillage (CT) is compared to no-tillage (NT) for rotations of corn-oat-meadow (COM), corn-soybean (CS), and continuous corn (CC).

to a depth of 225 mm. The rotation impact on soil N pools in the 0 to 75 mm depth increment was greater in the no-tillage than the conventional tillage.

Long-term maintenance of N pools is also affected by soil erosion losses. It is generally accepted that rotations promote the soil-building processes and reduce erosion losses. The impact of rotation on soil-building processes and reduce erosion is not as pronounced in conservation tillage systems as in conventional tillage systems. Van Doren et al. (1984) reported that erosivity of corn-soybean rotation averaged 45% greater than continuous corn when soil was conventionally tilled but found no indication that no-tillage was less effective in erosion control following soybean than following corn at equal residue cover.

Interaction with the Environment

Crop rotations have great appeal to those concerned with environmental quality. The extent to which rotations impact surface and water groundwater quality is related to their impact on chemical requirements and fate of applied chemicals relative to monoculture. While no cropping system is without problems, crop rotations can potentially reduce NO_3 movement to groundwater due to the following phenomena:

1. Crop N uptake patterns are potentially more synchronous with N-cycling patterns in soils in rotations than in certain monocultures, particularly corn. Certain crops in rotation and certain cover crops can act as catch crops for residual N. Crop residues from certain

crops, when properly placed in the soil, can immobilize soil N, delaying N release to a time more synchronous with the N-uptake patterns of the succeeding crop. Patriquin (1986) suggests that crop residues or cover crops prior to a bean legume will deplete inorganic nitrogen. If inorganic nitrogen is available to the bean, N_2 fixation will be inhibited. In his studies on biological husbandry, Patriquin (1986) suggests that the problem of N on the farm may not be one of quantity of N entering the system but the way it cycles within the system.

2. Nitrogen-use efficiency can improve in succeeding crops in rotation over that in monoculture. This may or may not be true of NUE_f (Fig. 3-1), but can be true of NUE partly because of increased N uptake.
3. Fertilizer N requirements may be reduced for crops grown in rotation. For example, this is well established in the case of alfalfa in rotation with corn (Blackmer, 1987; Hesterman et al., 1986; Triplett et al., 1969).
4. Water-use efficiency and efficient water use, if the distinction can be made, as discussed later, are potentially increased in rotation leading to greater and more efficient N uptake and reduced leaching losses.

On the negative side, WUE may decline in some rotations which may result in more N losses. However, the decline in WUE may not always be negative relative to crop yield. Substantial leaching of N can occur in some rotations, particularly those that include alfalfa (Bergstrom, 1987). In the wheat-fallow rotation, NO_3 accumulation during the fallow period may exceed subsequent crop requirements and result in N leaching below the root zone. Lamb et al. (1985) reported NO_3-N levels in November of the fallow year exceeding 160 kg ha^{-1} in the 0 to 1.2-m soil profile and that the amount of NO_3-N in the soil profile increased with fertilization. Janzen (1987) suggested that N fertilizer was needed to maintain sufficient levels of actively mineralizable N pools in soil where soil organic carbon and nitrogen losses have been accelerated by cultivation and frequency of summer fallow. Janzen (1987) states that regular additions of moderate levels of inorganic fertilizers would prevent future reliance on higher inputs of fertilizer resulting from depletion of organic nitrogen reserves. The accelerated loss in soil N is verified by the analysis of Juma and McGill (1986), who reported much lower steady-state levels of N in grain-fallow rotation than in continuous small grains at two locations in the Great Plains.

Schepers (1988) discusses more specific aspects of the environmental effects of rotations and cropping systems. The outlook for improvement of environmental quality with rotations is appealing. But, it may take some manipulation of the rotation and/or cropping systems to optimize their environmental benefits.

CROP ROTATION IMPACTS ON THE EFFICIENT USE OF WATER

While the efficient use of water has been thoroughly reviewed in recent years (Kozlowski, 1968; Taylor et al., 1983), the specific impact of crop rotations on water-use efficiency (WUE) has received little attention. The exception is the wheat-fallow rotation in the Great Plains region of the USA.

Terminology

Two terms are used in describing efficient use of water—"efficient water use" and "WUE". Tanner and Sinclair (1983) describe these phrases as intrinsically ambiguous in relation to crop production. Water-use efficiency is the biomass production per unit cropped area for a unit of water evaporated and transpired (ET). The traditional form of the relationship is

$$\mathrm{WUE} = \mathrm{Y}/\mathrm{ET}. \qquad [9]$$

This definition of WUE represents the basic component of the various WUE models discussed by Hanks (1983) and Tanner and Sinclair (1983). The phrase *efficient water use* is used to describe savings of water through a variety of means that can increase the total production per unit of water available. The statement of Tanner and Sinclair (1983) that water savings leading to efficient water use do not improve WUE is an important one since it points clearly to potential misuse and confusion in the use of these terms.

Water-Use Efficiency

Of the six models Viets (1962) proposed for the relation between ET and dry matter yield (Y), only two appear to exist in the field. Figure 3–4 reproduces the Y vs. ET and Y/ET vs. Y relationships proposed by Viets (1962). Under Viets' (1962) Model B, the ET vs. Y relationship is a linear function with an intercept (Fig. 3–4a), which gives a hyperbolic function of Y/ET vs. Y (Fig. 3–4b). Under Model B, WUE (Y/ET) increases as yield increases because water evaporated when plant growth is zero is wasted (Viets, 1962). Under Viets' (1962) Model D, ET is independent of Y after reasonably complete cover is attained (Fig. 3–4c). Water-use efficiency is a linear function of Y (Fig. 3–4d) whether yield is changed by season, irrigation regime, or fertilizer application (Viets, 1962).

If total dry matter rather than grain yield alone comprises Y, crop rotation under Model B has no impact on WUE if dry matter production is above the level indicated as the dashed line in Fig. 3–4c. Below this production level, crop rotations will increase or decrease WUE. Therefore, at today's level of crop production, no increase in WUE would be expected under Model B. Rotations might, however, be expected to increase WUE in low input agricultural systems. Under Model D, any change in Y results in a change

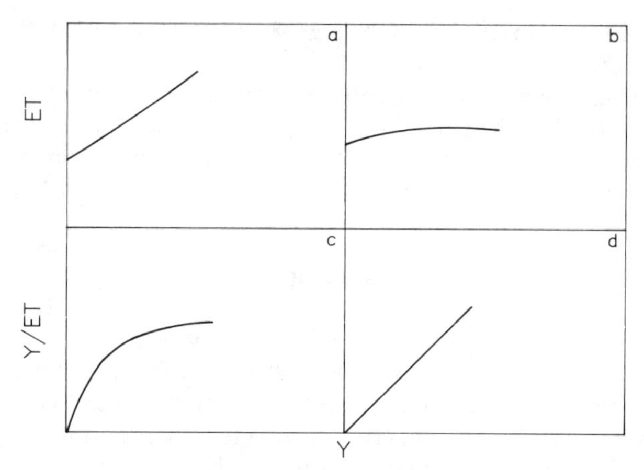

Fig. 3-4. Illustration of Viets' (1962) Model B (Fig. 3-4a and 3-4c) and Model D (Fig. 3-4b and 3-4d). As described by Viets (1962), under Model B, evapotranspiration (ET) is a linear function of biomass (Y) (Fig. 3-4a), with an intercept that gives a hyperbolic function of Y/ET (Fig. 3-4c). Under Model D, ET is independent of Y after reasonably complete cover is attained (Fig. 3-4b) and Y/ET is a linear function of Y (Fig. 3-4d).

in WUE. Therefore, rotations that improve yields would improve WUE and rotations that decrease yields would decrease WUE.

As to which model is operative in determining the impact of crop rotations on WUE is unclear. Hanks (1983) reported that the relationship between Y and ET is linear for a given crop, although crop variety and year appear to affect the relationship but not the linearity. Tanner and Sinclair (1983) also seem to support Model B in their perception of the invariant nature of WUE. Ritchie (1983) agrees that when management is fixed and water supply varies, WUE, when normalized for climate, is constant. However, he states that sufficient data where management is a variable are lacking in adequate detail to evaluate accurate biomass-ET curves. In support of Model D, Ritchie (1983) argued that data on the mean slope of the relationship between ET efficiency and yield as influenced by management for several crops (Table 3-4) clearly support the notion that management influences ET efficiency. Ritchie (1983) states, however, that "it seems probable that a difference in soil evaporation among various management treatments for the period of the season when leaf area index (LAI) <1 could account for most of the differences in measured ET efficiencies" (given in Table 3-4). Figure 3-5 gives the relationship between ET efficiency and biomass yield of wheat that Ritchie (1983) developed from data reported by Jensen and Sletten (1965) and Power et al. (1961). The connected solid lines represent changes in ET efficiency caused by fertilizers where ET was fixed (i.e., each solid line represents a different irrigation level), and conforms to principles of Viets' Model D (Ritchie, 1983). The dashed lines represent fixed fertility and variable water supply treatments and conform to principals of Viet's Model B at high yield levels. It would appear, therefore, that both models proposed by Viets (1962) may be operative but under different conditions.

Table 3-4. The mean slope of the relationship between ET efficiency and yield as influenced by management for several crops (Ritchie, 1983).

Crop	Management	Mean slope, m^{-1}	n	Reference
Wheat grain	N fertilization	2.9, ± 0.3	5	Bond et al., 1971
Wheat grain	N fertilization	1.2, ± 0.4	24	Jensen and Sletten, 1965
Sorghum grain	Population	2.7, ± 0.7	3	Olson, 1971
Sorghum grain	N fertilization	1.7, ± 0.1	22	Jensen and Sletten, 1965
Corn forage	N fertilization + population	2.7, ± 0.7	6	Carlson et al., 1959
Sorghum biomass	Population	3.8, ± 0.6	2	Olson, 1971
Corn biomass	Population	2.8	1	Olson, 1971
Forage sorghum	Population	3.2	1	Olson, 1971
	$\bar{x}$	2.6		

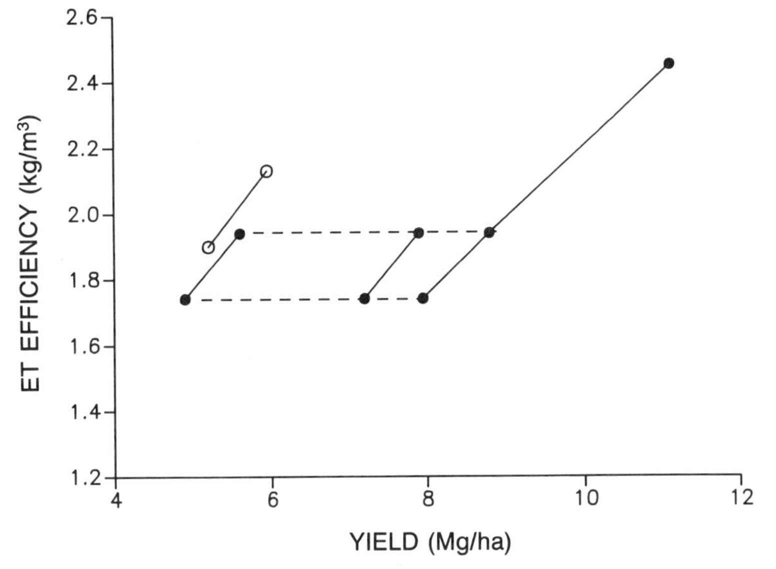

Fig. 3-5. Relationship between evapotranspiration (ET) efficiency and biomass yield of wheat developed by Ritchie (1983) from data reported by Jensen and Sletten (1965, open circles, ○) and Power et al. (1961, closed circles, ●). Each connected solid line represents the change in ET efficiency caused by fertilizer at one, two, and three irrigations. The dashed lines represent fixed fertility and variable water supply treatments. (Reproduced from *Limitations to Efficient Water Use in Crop Production.* 1983 p. 29-44. By permission of the ASA, CSSA, and SSSA, Inc., Madison, WI.)

Efficient Water Use

Recall that "EWU" refers to saving water from a fixed supply for crop use and differs from WUE. Crop rotations may impact water available for crop growth through its effect on processes controlling available water supply in soil, particularly infiltration and evaporation. Unger and Stewart (1983) cite increased aggregation and soil organic matter associated with rotation over monoculture as factors increasing infiltration and reducing runoff resulting in improved water relations and corresponding increased yields.

The concept of EWU is not well defined. We propose a set of definitions for water analogous to Bock's (1984) yield efficiency, recovery efficiency, and physiological efficiency for applied N. Yield efficiency is defined as the ratio of biomass (Y) to available water (W_a). Recovery efficiency is defined as the ratio of water used (W_u) to W_a and physiological efficiency as the ratio of Y to W_u. Efficient water use would be equivalent to yield efficiency and is the product of the recovery and physiological efficiencies (Eq. [10]).

$$\mathrm{EWU} = \frac{Y}{W_a} = \frac{W_u}{W_a} \times \frac{Y}{W_u}. \qquad [10]$$

W_a would constitute all water available to the plant during a defined period and could be calculated using mass balance principles. If W_u is defined in terms of ET (or T) then

$$\mathrm{EWU} = \frac{\mathrm{ET}}{W_a} \times \mathrm{WUE}. \qquad [11]$$

Equation [11] then defines EWU as a function of WUE and the recovery efficiency, ET/W_a. Under Viets' (1962) Model B, WUE is unaffected by rotation but ET increases when yield increases. Therefore, under Model B, rotations increase or decrease EWU when yields increase or decrease relative to monoculture and W_a is unaffected. Under Viets' (1962) Model D, WUE changes in proportion to yield because ET is unaffected. EWU changes in proportion to yield because ET is unaffected. Efficient water use changes in proportion to yield but modified to the extent that water availability (W_a) and ET are affected by rotation.

The value of these definitions is that EWU is expressed as a function of WUE modified by rotation's impact on water availability. In this way, rotation's impact on EWU can be positive or negative under either model B or D, depending on the rotation's influence on W_a and ET. This concept of EWU should be pursued and provides a means to quantify rotation effects on efficient water use.

SUMMARY AND CONCLUSION

Crop rotations have generally been viewed as beneficial. While beneficial, they have not always been viewed as economical. As Kurtz et al. (1984) suggested, N applications can, in many cases, replace the rotation effect as long as supplies remain unrestricted (economically or environmentally).

The impact of rotations on N-use efficiency was shown in this chapter to be dependent upon the definition of NUE, crop response in rotation to available N, and the time frame of consideration, in other words, the indirect or long-term rotation effects. We suggest that a true assessment of crop rotations impact on NUE will only be obtained when all N pools are considered and not solely through crop response to applied N. Therefore, any new studies directed at the question of the impact of crop rotation on NUE will need to better quantify the contribution and sources of all N pools to plant uptake within the agroecosystem.

Discussion of efficient water use in the literature suffers from ambiguous terminology. Under definitions proposed here, EWU is a function of WUE. Many studies show significant increase in yield or biomass when crops are grown in rotation. Rotations are often considered to improve water relations in soils. It is generally agreed that as biomass increases, water use increases. The two models proposed for the relationship between WUE and biomass seem to occur in the field. Under reasonable production levels, crop rotations would not increase WUE under Model B, while under Model D, any change in production would result in increased WUE. Based on our definition, rotations can affect EWU under either model.

There is a real lack of research that has determined either water and N use, or their interaction, as affected by crop rotation. Although we recognize the problems related to funding and methodology needed to conduct research that would answer questions regarding rotation impacts on NUE and WUE, efforts need to be directed towards this important topic.

REFERENCES

Abshahi, A., F.J. Hills, and F.E. Broadbent. 1984. Nitrogen utilization by wheat from residual sugarbeet fertilizer and soil incorporated sugarbeet tops. Agron. J. 76:954–958.

Arnon, I. 1972. Crop production in Dry Regions. Vol. 1. Backround and principles. Leonard Hill, London.

Baldock, J.O., R.L. Higgs, W.H. Paulson, J.A. Jackobs, and W.D. Shrader. 1981. Legume and mineral N effects on crop yields in several crop sequences in the Upper Mississippi Valley. Agron. J. 73:885-890.

Bergstrom, L. 1987. Nitrate leaching and drainage from annual and perennial crops in tile-drained plots and lysimeters. J. Environ. Qual. 16:11–18.

Bezdicek, D.F., R.F. Mulford, and B.H. Magee. 1974. Influence of organic nitrogen on soil nitrogen, nodulation, nitrogen fixation, and yield of soybeans. Soil Sci. Soc. Am. Proc. 38:268–272.

Blackmer, A.M. 1987. Losses of fertilizer N from soils. p. 51–61. *In* Conservation tillage on wet soils. Proc. Meet. Iowa and Minnesota Chapters. 18–19 February, Soil Conserv. Soc. of Am. Clear Lake, IA.

Bock, B.R. 1984. Efficient use of nitrogen in cropping systems. p. 273–294. *In* R.D. Hauck (ed.) Nitrogen in crop production. ASA, CSSA, and SSSA, Madison, WI.

Bolton, H., Jr., L.F. Elliott, R.I. Papendick, and D.F. Bezdicek. 1985. Soil microbial biomass and selected soil enzyme activities: Effect of fertilization and cropping practices. Soil Biol. Biochem. 17:297–302.

Bond, J.J., J.F. Power, and W.O. Willis. 1971. Soil water extraction by N-fertilized spring wheat. Agron. J. 63:280–283.

Burutolu, E.F.A. 1985. Influence of cultivation, cropping systems, and crop residues on the content and distribution of nitrogen forms in soils. Ph.D. diss. Michigan State Univ., East Lansing (Diss. Abstr. ADG 86-07054).

Carlson, C.W., J. Alessi, and R.H. Mickelson. 1959. Evapotranspiration and yield of corn as influenced by moisture level, nitrogen fertilization, and plant density. Soil Sci. Soc. Am. Proc. 23:242–245.

Curl, E.A. 1963. Control of plant diseases by crop rotation. Bot. Rev. 29:413–479.

Dick, W.A. 1984. Influence of long-term tillage and crop rotation combinations on soil enzyme activities. Soil Sci. Soc. Am. J. 48:569–574.

----, and D.M. Van Doren, Jr. 1985. Continuous tillage and rotation combinations effects on corn, soybean, and oat yields. Agron. J. 77:459–465.

----, ----, G.B. Triplett, Jr., and J.E. Henry. 1986a. Influence of long-term tillage and rotation combinations on crop yields and selected soil parameters. I. Results obtained for a Mollic Ochraqualf soil. Res. Bull. 1180. The Ohio State Univ., Ohio Agric. Res. Dev. Ctr., Wooster.

----, ----, ----, and ----. 1986b. Influence of long-term tillage and rotation combinations on crop yields and selected soil parameters. II. Results obtained for a Typic Fragiudalf soil. Res. Bull. 1181. The Ohio State Univ., Ohio Agric. Res. Dev. Ctr., Wooster.

Doster, D.H., D.R. Griffin, J.V. Mannering, and S.D. Parson. 1983. Economic returns from alternative corn and soybean tillage systems in Indiana. J. Soil Water Conserv. 38:504–508.

Hairston, J.E., J.O. Sanford, D.F. Pope, and D.A. Horneck. 1987. Soybean-wheat doublecropping: Implications from straw management and supplemental nitrogen. Agron. J. 79:281–286.

Hanks, R.J. 1983. Yield and water-use relationships: An overview. p. 393–412. *In* H.M. Taylor et al. (ed.) Limitation to efficient water use in crop production. ASA, CSSA, and SSSA, Madison, WI.

Hargrove, W.L., J.T. Touchton, and J.W. Johnson. 1983. Previous crop influence on fertilizer nitrogen requirements for double-cropped wheat. Agron. J. 75:855–859.

Hauck, R.D. (ed.) 1984. Nitrogen in crop production. ASA, CSSA, and SSSA, Madison, WI.

Hesterman, O.B., M.P. Russelle, C.C. Shaeffer, and G.H. Heichel. 1987. Nitrogen utilization from fertilizer and legume residues in legume-corn rotations. Agron. J. 79:726–731.

----, C.C. Sheaffer, D.K. Barnes, W.E. Lueschen, and J.H. Ford. 1986. Alfalfa dry matter and nitrogen production, and fertilizer response in legume-corn rotations. Agron. J. 78:19–23.

Hiltbold, A.E., R.M. Patterson, and R.B. Reed. 1985. Soil populations of *Rhizobium japonicum* in a cotton-corn-soybean rotation. Soil Sci. Soc. Am. J. 49:343–348.

Janzen, H.H. 1987. Effect of fertilizer on soil productivity in long-term spring wheat rotations. Can. J. Soil Sci. 67:165–174.

Jensen, M.E., and W.H. Sletten. 1965. Evapotranspiration and soil moisture-fertilizer interrelations with irrigated winter wheat in the southern High Plains. USDA-TAES Conserv. Res. Rep. 4. U.S. Gov. Print. Office, Washington, DC.

Juma, N.G., and W.B. McGill. 1986. Decomposition and nutrient cycling in agroecosystems. p. 74–136. *in* M.J. Mitchell and J.P. Nakas (ed.) Microfloral and faunal interactions in natural and agro-ecosystems. Nijhoff/Dr. W. Junk Pub., Boston.

Khan, S.U. 1971. Nitrogen fractions in a gray wooded soil as influenced by long-term cropping systems and fertilizers. Can. J. Soil Sci. 51:431–437.

Kozlowski, T.T. (ed.) 1968. Water deficits and plant growth. Academic Press, New York.

Kurtz, L.T., L.V. Boone, T.R. Peck, and R.G. Hoeft. 1984. Crop rotations for efficient nitrogen use. p. 295–306. *In* R.D. Hauck (ed.) Nitrogen in crop production. ASA, CSSA, and SSSA, Madison, WI.

Ladd, J.N., J.M. Oades, and M. Amato. 1981a. Microbial biomass formed from ^{14}C, ^{15}N-labelled plant material decomposing in soils in the field. Soil Biol. Biochem. 13:119–126.

----, ----, and ----. 1981b. Distribution and recovery of nitrogen from legume residues decomposing in soils sown to wheat in the field. Soil Biol. Biochem. 13:251–256.

----, and E.A. Paul. 1973. Changes in enzymic activity and distribution of acid-soluble, amino acid-nitrogen in soil during nitrogen immobilization and mineralization. Soil Biol. Biochem. 5:825–840.

Lamb, J.A., J.W. Doran, and G.A. Peterson. 1987. Nonsymbiotic dinitrogen fixation in no-till and conventional wheat-fallow systems. Soil Sci. Soc. Am. J. 51:356–361.

----, G.A. Peterson, and C.R. Fenster. 1985. Fallow nitrate accumulation in a wheat-fallow rotation as affected by tillage system. Soil Sci. Soc. Am. J. 49:1441–1446.

Levin, A., D.B. Beegle, and R.H. Fox. 1987. Effect of tillage on residual nitrogen availability from alfalfa to succeeding corn crops. Agron. J. 79:34–38.

Mannering, J.V., and D.R. Griffin. 1981. Value of crop rotations under various tillage systems. Purdue Univ. Coop. Ext. Serv. Pub. (Tillage) AY-230.

Meisinger, J.J. 1984. Evaluating plant-available nitrogen in soil-crop systems. p. 391–416. *In* R.D. Hauck (ed.) Nitrogen in crop production. ASA, CSSA, and SSSA, Madison, WI.

Olsen, R.J., R.F. Hensler, O.J. Attoe, S.A. Witael, and L.A. Peterson. 1970. Fertilizer nitrogen and crop rotation in relation to movement of nitrate nitrogen through soil profiles. Soil Sci. Soc. Am. Proc. 34:448–452.

Olson, T.C. 1971. Yield and water use by different populations of dryland corn, grain sorghum, and forage sorghum in the western corn belt. Agron. J. 63:104–106.

Patriquin, D.G. 1986. Biological husbandry and the "nitrogen problem." Biol. Agric. Hortic. 3:167–189.

Power, J.F. and J.W. Doran. 1988. Role of crop residue management in nitrogen cycling and use. p. 101–113. *In* Cropping strategies for efficient use of water and nitrogen. Spec. Pub. 51. ASA, CSSA, and SSSA, Madison, WI.

----, D.L. Grunes, and G.A. Reichmann. 1961. The influence of phosphorus fertilization and moisture on growth and nutrient absorption by spring wheat: I. Plant, growth, N uptake and moisture use. Soil Sci. Soc. Am. Proc. 25:207–210.

Powlson, D.S., P.C. Brookes, and B.T. Cristensen. 1987. Measurement of soil microbial biomass provides an early indication of changes in total soil organic matter due to straw incorporation. Soil Biol. Biochem. 19:159–164.

Radke, J., R.W. Andrews, R.R. Janke, and S.E. Peters. 1988. Low-input cropping systems and efficiency of water and nitrogen use. p. 193–218. *In* Cropping strategies for efficient use of water and nitrogen. Spec. Pub. 51. ASA, CSSA, and SSSA, Madison, WI.

Randall, G.W. 1987. Nitrogen management in conservation tillage systems. p. 86–92. *In* Conservation tillage on wet soils. Proc. Meet. Iowa and Minnesota Chapters, Soil Conserv. Soc. of Am., Clear Lake, IA. 18–19 February.

Rice, C.W., M.S. Smith, and R.L. Blevins. 1986. Soil nitrogen availability after long-term continuous no-tillage and conventional tillage corn production. Soil Sci. Soc. Am. J. 50:1206–1210.

Ritchie, J.T. 1983. Efficient water use in crop production: Discussion on the generality of relations between biomass production and evapotranspiration. p. 29–44. *In* H.M. Taylor et al. (ed.) Limitation to efficient water use in crop production. ASA, CSSA, and SSSA, Madison, WI.

Sanford, J.O., D.L. Myhre, and N.C. Merwine. 1973. Double cropping systems involving no-tillage and conventional tillage. Agron. J. 65:978–982.

Schepers, J.S. 1988. Role of cropping systems in environmental quality: Groundwater nitrogen. p. 167–178. *In* Cropping strategies for efficient use of water and nitrogen. Spec. Pub. 51. ASA, CSSA, and SSSA, Madison, WI.

Stevenson, F.J. 1982. Organic forms of soil nitrogen. *In* F.J. Stevenson (ed.) Nitrogen in agricultural soils. Agronomy 22:67–122.

Tanner, C.B., and T.R. Sinclair. 1983. Efficient water use in crop production: Research or re-search? p. 1–28. *In* H.M. Taylor et al. (ed.) Limitation to efficient water use in crop production. ASA, CSSA, and SSSA, Madison, WI.

Taylor, H.M., W.R. Jordan, and T.R. Sinclair (ed.) 1983. Limitation to efficient water use in crop production. ASA, CSSA, and SSSA, Madison WI.

Thurlow, D.L., J.H. Edwards, W. Gazaway, and J.T. Eason. 1985. Influence of tillage and crop rotation on soybean yields and cyst nematode population. p. 45–49. *In* W.L. Hargrove et al. (ed.) The rising hope of our land. Proc. 1985 Southern Region No-till Conf., Griffin, GA. 16–17 July.

Triplett, G.B., Jr. 1986. Crop management practices for surface tillage systems. p. 149–182. *In* M.A. Sprague and G.B. Triplett (ed.) No-tillage and surface tillage agriculture. The Tillage Revolution. John Wiley and Sons, New York.

----, F. Haghri, and D.M. Van Doren. 1969. Legumes supply nitrogen for no-tillage corn. Ohio Rep. 64:83–85.

Unger, P.W. 1968. Soil organic matter and nitrogen changes during 24 years of dryland wheat tillage and cropping practices. Soil Sci. Soc. Am. Proc. 32:427–429.

----, G.W. Langdale, and R.I. Papendick. 1988. Role of crop residues—Improving water conservation and use. p. 69–100. *In* Cropping strategies for efficient use of water and nitrogen. Spec. Pub. 51. ASA, CSSA, and SSSA, Madison, WI.

----, and B.A. Stewart. 1983. Soil management for efficient water use: An overview. p. 419–460. *In* H.M. Taylor et al. (ed.) Limitation to efficient water use in crop production. ASA, CSSA, and SSSA, Madison, WI.

USDA. 1973. Monoculture in agriculture: Extent, causes, and problems—Report of the task force on spatial heterogeneity in agricultural landscapes and enterprises. U.S. Gov. Print. Office, Washington, DC.

Van Doren. D.M., Jr., W.C. Moldenhauer, and G.B. Triplett, Jr. 1984. Influence of long-term tillage and crop rotation on water erosion. Soil Sci. Soc. Am. J. 48:636–640.

----, G.B. Triplett, Jr., and J.E. Henry. 1976. Influence of long term tillage, crop rotation, and soil type combinations on corn yield. Soil Sci. Soc. Am. J. 40:100–105.

Varco, J.J., W.W. Frye, and M.S. Smith. 1985. Nitrogen recovery by no-till and conventional till corn from cover crops. *In* W.L. Hargrove et al. (ed.) The rising hope of our land. Proc. 1985 Southern Region No-till Conf. Griffin, GA. 16–17 July.

Viets, F.G., Jr. 1962. Fertilizers and the efficient use of water. Adv. Agron. 14:223–264.

Wagger, M.G., D.E. Kissel, and S.J. Smith. 1985. Mineralization of nitrogen from nitrogen-15 labeled crop residues under field conditions. Soil Sci. Soc. Am. J. 49:1220–1226.

Yates, F. 1954. The analysis of experiments containing different crops. Biometrics 10:324–346.

Young, J.L., and J.L. Mortensen. 1958. Soil nitrogen complexes: I. Ohio Agric. Exp. Stn. Res. Circ. 61:1–18.

Zielke, R.C., and D.R. Christenson. 1986. Organic carbon and nitrogen changes in soil under selected cropping systems. Soil Sci. Soc. Am. J. 50:363–367.

4 Planting Geometries and the Efficient Use of Water and Nutrients[1]

R. E. Sojka
USDA-ARS
Kimberly, Idaho

D. L. Karlen
USDA-ARS,
Florence, South Carolina

E. J. Sadler
USDA-ARS
Florence, South Carolina

In nature and in the most primitive agricultural systems, seed distribution is broadcast across the landscape. Such a distribution results in nearly uniform spacial interaction of the developing phytomass. With the development of agrarian civilization has come an implement-dependent systemization of crop-planting patterns. This has brought about the planting of crops in uniform rows—from the drilling of small grains at inter-row spacings of 0.1 to 0.2 m and plant intra-row spacings of 1 to 5 cm, to the staking of horticultural and vine crops at 2- to 3-m inter-row spacings and typically 0.3- to 1-m intra-row spacings.

The implement dependence of agricultural cropping strategies has resulted in row cropping. The staple crops regarded as most suited to this approach are commonly called *row crops,* and this review will concentrate largely on how row crops interact with plant geometry, water, and nutrients to influence sustained productive capacity.

PRACTICAL AND HISTORICAL CONSIDERATIONS

The origins of particular row spacings can probably be traced back to implement development, beginning with animal-drawn implements. A mule

[1]Contribution of the USDA-ARS, Soil and Water Management Unit, Kimberly, ID 83341, and Coastal Plain Soil and Water Conserv. Res. Ctr., Florence, SC 29502.

Cropping Strategies for Efficient Use of Water and Nitrogen, Special Publication no. 51.

or a horse (*Equus caballus*) requires about 0.92 to 1.02 m (36–40 in.) of clearance to walk between planted rows with a one-row cultivator. The average 50-kW (80 hp) American tractor can adjust its wheel centers from 1.52 to 2.24 m (60–88 in.). At 1.52 m, the tractor can straddle two 0.76-m (30-in.) rows. At 1.83 m (72 in.), it can straddle three 0.61-m (24-in.) rows. At 2.24 m, the tractor can straddle four 0.56-m (22-in.) rows. At 2.04 m (80 in.), it could straddle four 0.51-m (20-in.) rows, but generally, 0.51 m is nearly equal to the narrowest tire width available. That would restrict tractor use to planting and perhaps one early cultivation. In most cases, farmers prefer to straddle even numbers of rows leaving an inter-row below the low-hanging tractor center and allowing operations to be done on even-row multiples that usually are more compatible with row-crop harvesting equipment.

Where land is furrow irrigated, again, row spacings closer than about 0.51 m are difficult to achieve. Closer-spaced furrows would be destroyed by tire traffic or would not be large enough to carry water the length of a typical field. Conversely, furrow spacings wider than 0.92 m are difficult to manage on single-rowed beds. This is because water will not move laterally (sub-across) from the irrigated furrow more than about 0.46 m by capillarity in a typical 12- to 24-h irrigation set, even on well-aggregated loamy soils. With sandy or clayey soils, the maximum manageable width would be even less. This limit on maximum furrow spacing is reinforced by the common on-farm practice of irrigating only alternate furrows to save labor and prevent overirrigation.

Wide row spacings continue to be used more extensively in the South than perhaps in any other part of the USA. This has probably occurred for several reasons. The Southern states, for socio-economic reasons, were the last to fully embrace agricultural mechanization (Healy, 1985). The use of draft animals remained common in the South through the late 1940s and early 1950s. Consequently, cotton (*Gossypium hirsutum* L.), corn (*Zea mays* L.), soybean [*Glycine max* (L.) Merr.] and tobacco (*Nicotiana tabacum* L.) remained in wide-row configurations well after extensive crop-breeding programs had gotten underway. Thus, these crops have been inadvertently bred for optimal response under wide-row configurations. The South's intense weed pressure also reinforced the need to cultivate late in the season, which is facilitated by wide rows.

The upsurgance of determinate soybean as a southern crop further played a role in establishing wide rows. If canopy closure occurs by flowering (a date dependent on maturity group and latitude, and usually near August, first at mid-southern latitudes), there is only a small impact of row spacing on yield (Beatty et al., 1982; Beaver and Johnson, 1981).

Also, southern states are typically dominated by ultisols with genetic and/or traffic induced hardpans. To promote rooting below these restricted layers, the use at planting of in-row subsoiling to a depth below the hardpan (as deep as 0.46 m) has become a common practice (Sojka et al., 1984b). An excessively close placement of subsoil shanks (the limit is about 0.6 m depending upon soil type, shank configuration, shank depth, and soil condition) results in failure of the shanks to act independently, thus plowing up large soil masses.

Among the most complicated agricultural machinery is self-propelled harvesting equipment. Implements such as grain combines, corn pickers, cotton pickers, cotton strippers, potato diggers, and sugarbeet lifters, have been engineered to the prevailing row configurations of their intended crops, which have come about more from historical than technical considerations. They are usually of fixed configuration or at best only minimally adjustable. The adjustments, when available, are usually intended as one-time set-up adjustments and not for multiple adjustments during a single harvest or between crops in a given year. Since it is seldom economically feasible to acquire multiple fleets of farm equipment tailored to more than one canopy configuration, most farmers choose a compromise row spacing for all crops in their rotation.

INCREASING CANOPY DENSITY

The historical and practical considerations notwithstanding, there has been a longstanding interest in increasing production by covering the ground earlier in the season with foliage from more and/or closer-spaced plants and plant rows (Bryant et al., 1940; Jordan et al., 1950; Mooers, 1910; Morrow, 1890; Nelson, 1931; Painter and Leamer, 1953; Probst, 1945; Reynolds, 1926; Wiggams, 1939). The simplest components of planting geometry that can be manipulated to affect row-crop performance are inter- and intra-row plant spacing. A broad generalization of theory and historical results across species and environments is that yield increases as canopy density, fertility, and soil water availability simultaneously increase until an optimum density is achieved, beyond which higher density reduces yield due to competition, lodging, etc., regardless of further increases in fertility or water availability (Chandler, 1969; Fontes and Ohlrogge, 1972; Lehman and Lambert, 1960; Weber et al., 1966). A further theoretical production limit exists under the constraints of photosynthetic efficiency of the respective C3 or C4 pathways. Morphological expression is an additional limitation resulting from elongation of light-restricted plants as canopy densities increase, eventually predisposing the crop to lodging. Some of the advantages and disadvantages associated with wide vs. narrow row planting patterns are listed in Table 4–1.

SOYBEAN

Developmental Effects

Perhaps the most abundant literature on planting geometry effects on growth and performance is for soybean. The various agronomic considerations listed in Table 4–1 assume comparing canopy geometries at a fixed plant population. Many negative aspects of either wide or narrow rows can be

Table 4-1. Agronomic considerations for wide or narrow row spacings at a fixed population per hectare.

Consideration	Row space result	
	Wide	Narrow
1. Planters per tool bar	Fewer	More
2. Passes/ha	Fewer	More
3. Fuel consumed/planted ha	Less	More
4. Fuel consumed/ha for in-row subsoilers	Less	More
5. Seed/ha for equal emergence	Less	More
6. Amount of banded pesticide/ha	Less	More
7. Yield loss to individual seed skips	Less	More
8. Yield loss to skipped row segments	More	Less
9. Time interval to canopy closure	Longer	Shorter
10. Inter-row shading	Less	More
11. Competition against weeds	Later	Earlier
12. Inter-row rooting	Less	More
13. Furrow irrigation set time per unit water infiltrated	Longer	Shorter
14. Water use	Management and environment dependent	
15. Stalk size	Smaller	Larger
16. Plant height	Taller	Shorter
17. Height to lowest pod in legumes	Higher	Lower
18. Spatial dependency of nutrient and water extraction	Greater	Less
19. Yield/photoassimilate efficiency of indeterminates	Less	Greater
20. Extraction efficiency of broadcast fertilizer	Less	Greater
21. Suitability to placed fertilizer	Greater	Less
22. Performance of platform-cutter combines	Reduced	Enhanced
23. Performance of row-crop header combines	Enhanced	Reduced

mitigated in soybean to some extent by increasing or decreasing the seeding rate. For example, farmers usually note that less seed per hectare is required to obtain a given per hectare stand in wide rows. The closer proximity of seeds to one another in the row enhances the emergence potential of neighboring seeds through interaction of the zones of soil-active forces exerted by each emerging seedling (a buddy effect).

Phenological expression in soybean also seems to influence the interaction of population and environment. Determinacy (limitation of vegetative growth and setting of potential reproductive positions during a fixed limited time period) in full-season plantings largely defeats the yield-enhancement potential of more rapid canopy closure (Beaver and Johnson, 1981). As determinate soybean are planted closer to the flowering date in late-season plantings, the effect mimics the mechanism of indeterminate soybean, increasing yield with narrower rows (Beatty et al., 1982; Beaver and Johnson, 1981; Boquet et al., 1982; Caviness, 1966; Caviness and Smith, 1959; Chan et al., 1980; Williams et al., 1970). This explains why narrow rows greatly increase yields of northern-latitude, indeterminate (low maturity group. . .00-III) soybean in full-season plantings (Cooper, 1977; Costa et al., 1980; Leffel and Barber, 1961; Ryder and Beuerlein, 1979; Safo-Kantanka and Lawson, 1980; Taylor et al., 1982, Weber et al., 1966; Wilcox, 1974) but have little effect on southern-latitude, determinate (high maturity group. . .IV-X) soybean in full-season plantings (Beatty et al., 1982; Camper and Smith, 1958; Carter and Boerma, 1979; Caviness, 1966; Caviness and Smith, 1959; Frans, 1959; Hartwig, 1954, 1957; Parker et al., 1981; Smith, 1952).

The effect of soybean intra-row plant spacing seems less important than the effect of inter-row spacing. This may result largely from the soybean's great capacity to morphologically compensate for changes in competition (Hinson and Hanson, 1962; Ramseur et al., 1984). Hartwig (1957) observed maximum yields of determinate soybean at 0.9-m inter-row spacing with 4-cm intra-row spacing of seeds. Intra-row spacings of 4 cm up to 46 cm produced nearly the same yields (Basnet et al., 1974; Donovan et al., 1963; Hoggard et al., 1978; Johnson and Harris, 1967; Lueschen and Hicks, 1977; Probst, 1945, Ramseur et al., 1984). As a result, significant skips in the row have nearly no effect on soybean yield (Caviness, 1961, 1966; Stivers and Swearingin, 1980). Increasing soybean plant densities usually result in increased plant height, height of the lowest pod, and lodging potential (Beatty et al., 1982; Beaver and Johnson, 1981; Cooper, 1971; Wilcox, 1974).

Radiation Interception

A major motivation for changing plant geometry is to improve light interception (Loomis and Williams, 1969; Mitchell, 1970; Pendleton, 1966; Shaw and Weber, 1967). Wide-row soybean culture results in a slower increase in leaf area index (LAI) than for narrow-row culture (Weber et al., 1966). Also, as seen in Fig. 4–1, radiation interception at corresponding LAI is less efficient for wide rows (Hicks et al., 1969; Taylor et al., 1982; Shibles and Weber, 1966). As a result of earlier canopy closure, a densely shaded canopy floor provides better weed control under narrow rows (Burnside et al., 1964; Burnside and Colville, 1964; Dougherty, 1969; Felton, 1976; Frans, 1959; Howe and Oliver, 1987; Kust and Smith, 1969; Peters, 1965; Peters

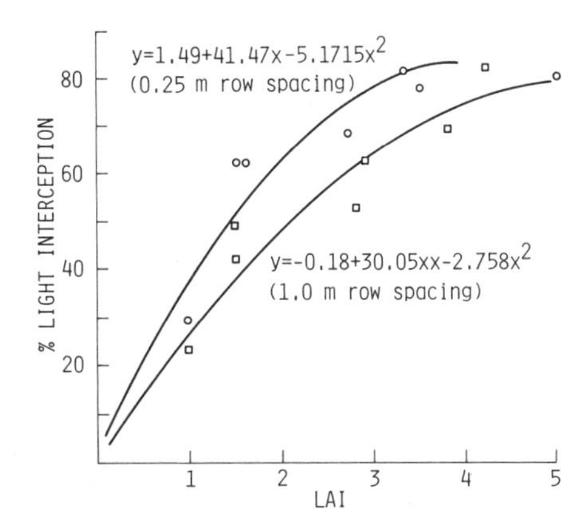

Fig. 4–1. Percentage of light interception in a soybean canopy as a function of leaf area index during canopy development. Circles are for 0.25-m row spacings; squares are for 1.0-m row spacings. From Taylor et al. (1982).

Table 4-2. Soybean response to irradiation of leveas with red (R) or far-red (FR) light. From Balatti and Montaldi (1986).

Plant property	R	FR
Foliar area, m^2	0.006572a*	0.006322a
No. nodules	7a	27b
Fresh wt. of nodules, g	0.00891a	0.04290b
Dry wt. of nodules, g	0.00139a	0.00874b
Leaf dry wt., g	0.26640a	0.28090a
Nitrogen, g/kg	28.8a	30.0a
Total N, g/plant	0.00767a	0.00842a

* Numbers in the same row followed by the same letter differ at $P = 0.05$ by the Duncan's multiple range test.

et al., 1965; Prasad et al., 1985a, b; Wax and Pendleton, 1968). The yield benefits of narrow rows can be lost entirely, however, if initial weed control is unsatisfactory, because of the inability to cultivate (Nave and Cooper, 1974; Wax et al., 1977). Several new highly effective over-the-top grass and broadleaf broad-spectrum herbicides are now available and yield loss need not result from inadequate early weed control. Management requirements, however, are more demanding when depending on these materials (Gebhardt and Minor, 1983), and economics may not be as favorable.

The quantity and pathway of radiation intercepted by the plant canopy are both affected by the canopy geometry (Holmes, 1981). This produces numerous environmental alterations including temperature distributions within the soil and canopy, foliar distribution of photosynthetically active radiation (PAR), and changes in canopy light quality.

Kasperbauer et al. (1984) demonstrated that canopy spectral composition changed in rows depending on their compass orientation, varying the exposure of the growing plant parts to far red light throughout the day and at extinction in the evening. In addition to the well-known effects of light spectral quality on flowering and shoot morphology, light-quality effects have also been demonstrated on nodulation and rooting (Balatti and Montaldi, 1983, 1986; Kasperbauer et al., 1984; Lie, 1969; Malik et al., 1982). Greater soybean rooting and nodulation (Table 4-2) have been associated with as little as 5 min of red/far-red exposure at photoperiod extinction (Balatti and Montaldi, 1983, 1986; Kasperbauer et al., 1984). Top dry weight was unaffected in these studies, and Balatti and Montaldi (1983, 1986) observed no differences in shoot N concentration or accumulation. Nearly opposite effects were observed in Lie's (1969) work with *Pisum sativum* and *Phaseolus vulgaris* which are long-day and day-neutral plants, respectively, whereas soybean is a short-day plant. These experiments suggest that environments that significantly alter light spectral composition, particularly at daylight extinction, may significantly impact root development and activity, specifically governed by each species' phytochrome adaptations. Further research is needed to determine the extent to which on-farm management of row-spacing, row-orientation, and plant population can beneficially manipulate these interactions.

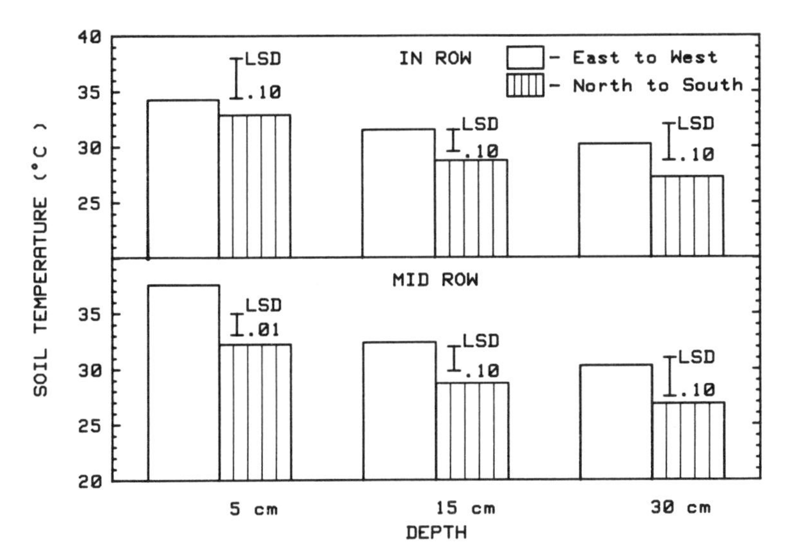

Fig. 4-2. The effect of row orientation on midday soil temperatures beneath a partially closed soybean canopy. From Hunt et al. (1985).

An interaction of soybean cultivars to *rhizobial* inoculants, row orientation and irrigation was observed by Hunt et al. (1985). They interpreted the responses as being largely soil temperature (Fig. 4-2) effects on cultivar × strain interactions. No significant differences in N concentration or accumulation occurred. In 2 of the 3 yr studied, however, irrigated North-South (N-S) oriented rows outyielded East-West (E-W) oriented rows across all strains with no consistent response in the 3rd yr. Mean soil temperatures prior to canopy closure were 3.2 °C higher for E-W oriented rows. Munevar and Wollum (1981a, b, 1982) had determined in the laboratory that for the strains later used in Hunt et al's (1985) field studies that the temperature differences between row orientations was sufficiently large to cause growth and host responses of free-living rhizobia of the magnitude and direction observed by Hunt et al. (1985). These temperature differentials were sufficient to assume an impact on root metabolism. Higher root-respiration of E-W oriented rows may have accounted for decreased yields. In addition, there may have been differences in nutrient availability related to ion activity and root absorptive capacity.

Water Use

Soybean water relations in general have been studied by a large number of researchers, but few have studied water relations as affected by planting geometry. Doss and Thurlow (1974) compared soybean performance in wide and narrow rows with high and low populations under irrigated and nonirrigated conditions. They found no influence on yield of inter- or intra-row spacing under irrigated conditions but found higher yields under low populations in the nonirrigated treatment. In earlier studies, Timmons et al. (1967)

and Peters and Johnson (1960) found the highest water-use efficiency (yield/evapotranspiration) in their narrow row treatments. Peters and Johnson (1960) determined that doubling plant population by decreasing inter-row spacing by one-half had a doubling effect on transpiration from flowering to maturity. In the study of Timmons et al. (1967), however, neither row spacing nor plant population affected evapotranspiration. Shibles et al. (1975) determined that until attainment of complete ground cover, transpiration varies in soybean canopies as a function of LAI but after complete canopy closure the aerial environment alone regulates ET.

Alessi and Power (1982) examined planting geometry effect on soybean water use from a dryland perspective. In 2 of 4 yr, they found that soybean yields were lowest, water use was highest, and for 3 of 4 yr, water-use efficiency was least for the narrowest row width they studied. Their data showed greater water use in narrow rows before flowering. In dryland situations or drought years, this early depletion caused by earlier canopy closure and increased ET (Reicosky et al., 1982b) leaves insufficient soil water for completing reproductive growth. This can reverse the usual expectation (especially for northern, indeterminate soybean) of increased yields with narrower rows, and underscores the importance of adequate water availability. Taylor (1980) found similar results. In the two driest years of a 3-yr study, there was no yield advantage for narrow rows, but a 17% increase occurred when seasonal water was adequate. In the dry years, wide-row plants grew taller, set more pods, and maintained higher leaf water potential (Ψ_L) than narrow-row plants. The amount of water conserved by wide rows in dry years, however, was evidently sufficient to maintain the early season biomass advantage, but insufficient to support an enhanced yield potential.

A similar response was reported by Campbell et al. (1984) for southern determinate soybean grown in a tillage study. In a year in which postflowering drought occurred, they found all treatments favoring water conservation (these included wide rows, maintenance of surface residues, early maturity groups, or delayed vegetative development) increased yields.

Although Taylor (1980) measured higher Ψ_L in wide-row soybean, other workers have had less success in determining row space-related differences in plant water status. Sojka and Parsons (1983), Sojka et al. (1984a), and Reicosky et al. (1985), reported greater differences related to cultivar than to row spacing and had difficulty delineating irrigation-related effects on plant water status. Sojka and Parsons (1983) and Sojka et al. (1984a) observed no significant row space-related differences in leaf temperature, Ψ_L, parallel leaf-diffusive resistance, vapor-pressure deficit, or leaf-air temperature differential. Reicosky et al. (1985) could not discriminate differences between Ψ_L in wide vs. narrow rows and found that Ψ_L differences between Ψ_L differences with irrigation were only significant under severe water stress.

Reicosky et al. (1982b, 1985) found a slight increase in ET of 0.15- or 0.25-m row spaced irrigated soybean compared to wider-spaced rows. They believed this increase in ET was related to the higher early season LAI and light interception of the narrow rows. They also showed a slightly greater root-length density in the narrow rows. Mason et al. (1982) and Taylor et al. (1982) also saw increased radiation interception, LAI, shoot to root ratio, and yield in narrow rows. Mason et al. (1982) further determined that the narrow-row treatments produced 49% more roots per hectare and 52% more roots per unit leaf area at identical plant populations. Despite the different root densities, there were no consistent differences for nonirrigated treatments in Ψ_L, soil temperatures, or water use over time or depth. Reicosky et al. (1982a) determined that the relationship between ET and Ψ_L was similar for both wide and narrow rows. This relationship was much more affected by application or absence of irrigation than by row spacing. There was, nonetheless, greater hysteresis in Ψ_L-ET diurnal curves for the non-irrigated narrow rows, which they interpreted as indicating greater early season water extraction.

Nutrient Use

The importance of adequate N availability (either as applied N fertilizer or as fixed N) to the effectiveness of narrow-row soybean culture was demonstrated by Cooper and Jeffers (1984). Nitrogen uptake rate and soil N depletion can be expected to occur more rapidly under narrow-row culture of soybean because more nearly equidistant spacing results in increased density of roots (Bohm, 1977; Taylor, 1980) as shown in Fig. 4-3, particularly with the common production practice of increasing plant populations in narrow-row systems.

Although N accumulation or seed yield did not increase, Bello et al. (1980) observed higher N_2-fixation rates (acetylene reduction) for higher plant populations and narrower row spacings. Maximum nodulation and N_2 fixation had earlier been shown to depend on adequate fertilization with P, K, and Ca (deMooy and Pesek, 1966; Fellers, 1918; Heltz and Whiting, 1928; Ludecke, 1941; Poschenrieder et al., 1940; Wilson, 1917). The rates required for this effect are higher than normal commercial fertilizer-application rates. Coupled with the recognized requirement for enhanced soil fertility in denser canopies, a heretofore unrecognized need for higher soil test values for narrow-row culture of soybean may exist.

Apart from the studies of soybean N production and uptake discussed above, there have been only a limited number of examinations of the interaction of soybean planting geometry and plant water status on nutrient accumulation. Bennie et al. (1982) found that Iowa soybean grown in 1.0-m row widths, regardless of irrigation, accumulated N, P, K, Ca, Mg, Na, Cu, Zn, Mn, Fe, B, and Al at a faster rate during the linear stage of nutrient uptake (between 49 and 91 d after planting) than those grown in 0.25-m row width. Concentrations of Mn were higher in 1.0-m row plants regardless of irrigation and Fe was higher in irrigated 1.0-m row plants. Sojka et al. (1984a) observed greater soil K^+ depletion in narrow rows but also found that depletion relative to row geometry was dependent on irrigation regime (Table 4-3), with greater depletion in the irrigated treatment. Furthermore, they found greater between-row depletion of K, Ca, and Mg. This may have indicated a concentrating effect within the row or that severe leaching effects oc-

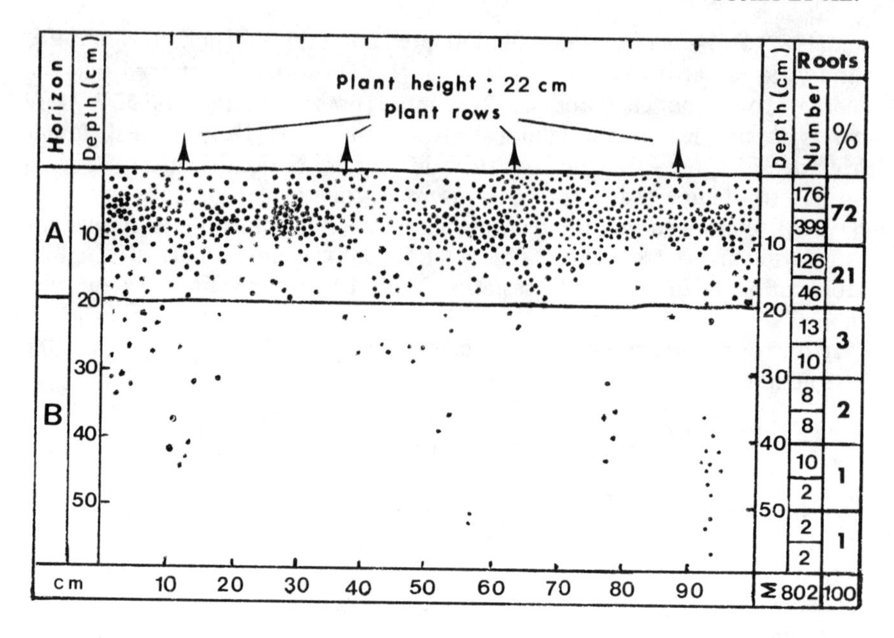

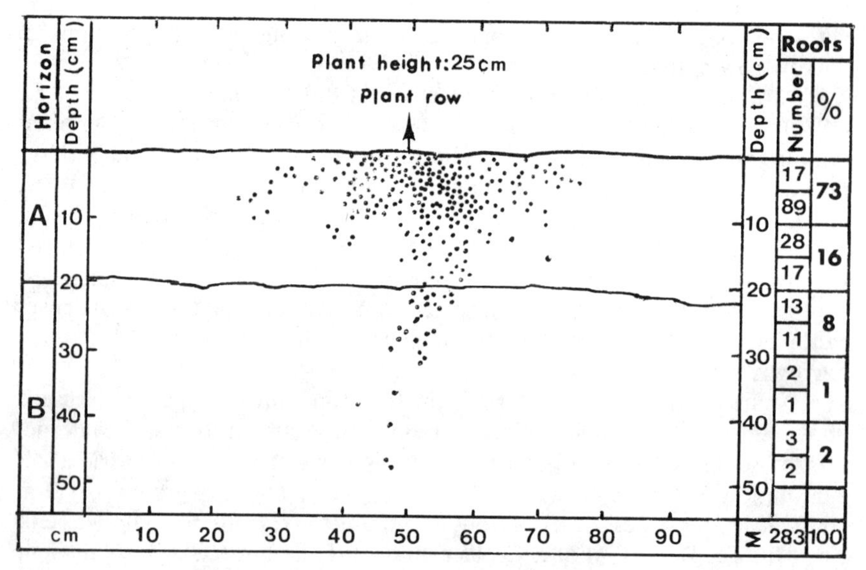

Fig. 4-3. Comparison of profile soybean root distribution 44 d after planting in 0.25-m vs. 1.00-m rows. Each dot represents approximately 5.0-mm root length on the exposed soil profile. From Bohm (1977).

curred between wide rows associated with high rainfall and sandy soils of the southeastern Coastal Plains. Mason et al. (1980) presented a detailed comprehensive summary of seasonal plant nutrient concentrations and accumulations by plant part. In their study, row spacing had little effect on plant

Table 4-3. Postharvest Mehlich I extractable K, Ca, and Mg as influenced by soybean row spacing, sampling depth, sampling position, and year. From Sojka et al. (1984).

Sampling depth	Row spacing	Sampling position	1980			1980		
			K	Ca	Mg	K	Ca	Mg
m			g/Mg					
0-0.15	0.50	M†	49	410	76	72	435	122
0-0.15	0.50	R	57	385	72	68	445	124
0-0.15	1.00	M	38	355	64	64	410	101
0-0.15	1.00	R	75	430	84	82	449	108
0.15-0.30	0.50	M	29	123	24	33	143	36
0.15-0.30	0.50	R	31	169	26	37	218	57
0.15-0.30	1.00	M	34	116	25	41	160	35
0.15-0.30	1.00	R	36	164	28	38	207	55
0.30-0.60	0.50	M	42	152	43	46	250	69
0.30-0.60	0.50	R	43	152	44	40	198	58
0.30-0.60	1.00	M	58	221	64	56	223	66
0.30-0.60	1.00	R	55	193	57	49	186	58
0.60-0.90	0.50	M	30	230	53	28	273	73
0.60-0.90	0.50	R	31	218	55	32	292	77
0.60-0.90	1.00	M	30	245	64	36	250	63
0.60-0.90	1.00	R	36	252	64	34	276	66
LSD (0.05)			10	116	15	10	77	18

† Sampling position M indicates between soybean rows while R indicates within the rows.

nutrient concentrations even though plants grown in wide rows were greener in color. Nodulation differences could not be detected. Because of greater biomass, there was greater per hectare elemental accumulation of all elements except Zn on a whole plant basis in wide rows. There were, however, higher concentrations in narrow rows of: pod wall P, stem and pod wall K, and whole plant B and Zn. Higher concentration of whole plant Mn and Fe occurred in wide rows.

CORN AND SORGHUM

The principles governing the effect of planting geometry on corn and sorghum [*Sorghum bicolor* (L.) Moench] are similar to those for soybean but are affected by the absence of the N_2-fixing process, and different root and shoot growth habits. The topic was reviewed by Duncan (1969); Dungan et al. (1958); Hinkle and Garrett (1961); Pendleton (1966); Stringfield (1962); and has been dealth with in varying degrees by other authors more recently as well (Blad, 1983; Cardwell, 1982; Waldren, 1983). In general, as population increases and row spacing decreases, water and nutrient availability plus overall management intensity must increase to optimize yields (Brown and Shroder, 1959; Grimes and Musick, 1959). Furthermore, Stringfield (1962) and Pendleton (1966) noted that these inputs must be expected to intensify even further as varieties are continually improved.

Conflicting experimental results have been accumulated related to the effects of row space and water use. Significant increases in corn and sorghum yields have been frequently reported for narrow rows and/or increased populations with good management (Andrews and Peek, 1971; Brown et al., 1970; Camp et al., 1985; Colville and Furrer, 1964; Colville, 1966; Downey, 1971; Duncan, 1958; Hoff and Mederski, 1960; Karlen et al., 1987; Kohnke and Miles, 1951; Lang et al., 1956; Larson and Hanway, 1977; Laude et al., 1955; Lutz et al., 1971; Porter et al., 1960; Sentz, 1965; Stickler, 1964; Stivers et al., 1971; Wooley et al., 1962; Yao and Shaw, 1964a.)

Radiation Interception

As with soybean, the increased yield of narrower rows appears to be the result of more efficient interception of PAR, either by increased LAI or more uniform spatial distribution, especially early in the season. The increased yield generally results in higher water-use effiency without significantly affecting seasonal water use (Aubertin and Peters, 1961; Colville, 1968; Denmead et al., 1962; Duncan, 1972; Knipmeyer et al., 1962; Pendleton et al., 1966; Peters and Russell, 1959; Tanner, 1957; Tanner et al., 1960; Timmons et al., 1966; Yao and Shaw, 1964a, b).

Corn may be more sensitive to competition than soybean, and for that reason stand uniformity can significantly affect the success of any planting geometry. Theoretically, equidistant spacing is optimum (Aldrich et al., 1976; Shubeck and Young, 1970). The need to ensure stand uniformity increases as all other management factors become more intensive, particularly with higher populations (Duncan, 1969).

Hill planting and other techniques resulting in uneven stands have generally yielded less than uniformly spaced stands (Colville and Furrer, 1964; Mock and Heghin, 1976; Pendleton, 1966; Waldren, 1983). This is due to the corn plant's tendency toward early adjustments to any inter- or intraspecies competition for light, water, or nutrients (Donald, 1958; Duncan, 1969; Hozumi et al., 1955; Waldren, 1983; Yoda et al., 1957). The tendency of closely spaced, shaded corn plants to elongate more rapidly than sunlit ones was noted by Hozumi et al. (1955). They further noted a lower phytomass accumulation in the shaded individuals. If an individual falls too far behind its neighbor it will continue to grow, using water, nutrients, and light at the expense of its neighbors, but itself remain barren. Karlen and Sojka (1985) referred to such unsuccessful individuals as "corn weeds."

As mentioned earlier, acceptance of new practices by farmers is increased when conventional equipment can be used. The objective of more uniform crop spacing, with correspondingly more uniform LAI distribution and increased yield, can be accomplished through reducing the row spacing for a given population. However, narrow rows generally require re-tooling of planters and cultivators, and purchase of narrow-row headers in the case of corn. The latter costs can be avoided in many cases by using twin rows, which are simply a pair of closely spaced rows centered on the conventional wide spacing. Karlen and Camp (1985), and Karlen et al. (1985) showed grain yields 5 to 8% higher in twin over single-row culture at constant populations. Earlier, corn silage yields had been shown to increase in twin rows (Bryant and Blaser, 1968; Washko and Kjelgaard, 1966). Twin rows were found to have similar cultural advantages for determinate soybean, but not to have significant yield advantages (Sojka, 1985, unpublished data).

The interactions of light intensity and quality under varying corn canopies was examined by Karlen et al. (1987) and Karlen and Kasperbauer (1988). They observed minimal difference between E-W or N-S oriented rows or variations in row configurations. Their 0.19–0.57–0.19-m twin rows yielded 5 to 10% better than 0.96-m single rows but not better than 0.76-m single rows. There were not large spectral variations due to spacing or orientation within the canopy and unlike soybean (Kasperbauer et al., 1984) physiological responses were not strongly tied to red/far-red exposure regimes. Earlier, Yao and Shaw (1964a) also failed to see significant differences in performance related to row orientation.

Water Use

The relative amount of water use between wide or narrow rows has been explained by some researchers as being dependent on the amount of surface soil water. Dry soil (stage three evaporation) resists vapor transfer and is an effective supplier of sensible heat to the wide-row plants, increasing their transpirative demand per unit leaf area (Chin Choy and Kanemasu, 1974; Chin Choy et al., 1977; Kanemasu and Arkin, 1974; McCauley et al., 1978; Yao and Shaw, 1964a). In addition, advection between wide rows can significantly increase the inter-row sensible heat balance, resulting in considerably higher ET rates (Blad, 1983; Chin Choy and Kanemasu, 1974; Hanks et al., 1971). Hanks et al. (1971) documented the significance of row advection as an energy source for sorghum ET; 21% of the dryland ET energy requirement, and 64% of the irrigated, originated from advection between 1-m rows in Akron, CO. Chin Choy and Kanemasu (1974) attributed the 10% higher ET from wide than from narrow-row sorghum to row advection early, and large-scale advection late in the season at Manhattan, KS.

In situations where the soil surface is frequently rehydrated by rain or irrigation (soil surface remains in Stage 1 or 2 evaporation), evaporation from the soil surface eliminates the source of sensible heat and ET is similar between the two canopies. Shading in the narrow-row canopy reduces the sensible heat load enough to slightly reduce ET in some instances. Under irrigation, then, one would expect increased weight of grain per unit ET (WUE) in narrow rows due to increase in grain yield resulting from significantly increased efficiency of light interception (Waldren, 1983). Forage yield may not increase, however (Cummins and Dobson, 1973). These results are contrary to the row space-related water-use patterns reported for soybean, but the contradicting results probably relate to experimental artifacts as discussed below.

Interpretation of these differences in water use between narrow and wide-row spacings is complicated by at least two factors. First, results vary depending upon climate and irrigation, and second, some experiments have been conducted using constant populations per unit ground area and others with constant plant numbers per unit row length. In these latter experiments, there exist both row-spacing and population variables, the interaction of which has not been completely described. Timmons et al. (1966) showed optimum populations for wide-row corn in the northwestern Corn Belt to increase in years with more water available during the season. In these tests, there was no clear population effect on seasonal water use, although the use generally trended upward with increased population. Contrastingly, there are data showing significant row spacing effects at constant populations. Yao and Shaw (1964a, b) showed higher water use for wide-row corn (0.53-, 0.81-, and 1.06-m spacings) in Iowa. Olson (1971) found no differences in water use for corn grown in 1.02-m rows at 35 000 plants/ha, in 0.51-m rows at 70 000 plants/ha, or in 0.51-m rows at 45 000 plants/ha.

More data exist for sorghum. Bond et al. (1964) tested 1.01- and 0.51-m row spacings, 4.4- and 8.8-kg/ha seeding rates (populations of about 45 000 and 90 000 plants/ha), and four initial soil-moisture levels in the southern High Plains. There was no significant effect of either row spacing or seeding rate on seasonal water use. However, narrower rows and higher populations shifted water use earlier in the season. Therefore, the fact that water is the limiting factor in the climate may have equalized seasonal water use. Plaut et al. (1969) studied irrigation timing on sorghum yield and water use, using a constant within-row spacing. In 1964, the apparently wetter year, 0.45-m row spacing yielded higher than 0.70-m spacing, and also used slightly more water in two of three comparisons. In the 2nd yr, with lower ET, the yield relationship reversed, and ET values were nearly the same for both row spacings. Olson (1971) showed no significant effect of row spacing on water use of either forage or grain sorghum at a constant within-row spacing in South Dakota. However, the narrow-row spacing, which had twice the population, had numerically higher water use in all years. Chin Choy and Kanemasu (1974) reported energy balances for wide (0.92-m) and narrow (0.46-m) row sorghum at a constant 12 plants/m of row. Seasonal ET was 10% higher from the wide-row sorghum, in spite of the higher population in narrow rows.

In a recent study, Steiner (1986) reported that in a dry year narrow rows and higher populations increased seasonal ET by 7 and 9% (Table 4–4), respectively, mostly due to increased prereproductive ET. Row direction did not affect water use or yield although the dry matter to ET ratio and light interception was higher in the narrow-row crop. In a 2nd yr of the study, there was more rain, but narrow-row ET was still higher between emergence and anthesis. Intensive observation in the 2nd yr (Steiner, 1987) indicated that net radiation was 5% higher over wide compared to narrow rows and E-W rows had 14% higher net radiation than N-S rows. Higher leaf temperatures were associated with higher populations caused by greater depletion of plant available water.

It is apparent that no single summary statement can be made to include all of the foregoing results. In general, higher populations and narrower row spacings used slightly more water in some experiments, or used water earlier in the season, thus exhausting the supply in water-limiting environments. In other experiments, the water-use effects were reversed, with more ET from

Table 4-4. Row spacing and population effects on evapotranspiration (ET), yield, and water-use efficiency (WUE) of dryland grain sorghum. From Steiner (1986).

Treatment	Seasonal ET	Vegetative ET	Grain-fill ET	ET during grain-fill	Total dry matter	Grain dry matter	Harvest index	WUE total	WUE grain
	— m —			%	— Mg/ha —			— kg/m³ —	
				1983					
Spacing, m									
0.38	0.200**	0.175**	0.025*	12**	5.92	2.03	0.34	2.96*	1.02*
0.76	0.184	0.151	0.033	18	5.92	2.13	0.36	3.22	1.16
Population									
High	0.199**	0.174**	0.025	13*	6.10*	1.64**	0.27**	3.08	0.83**
Medium	0.194	0.166	0.028	14	6.22	2.31	0.37	3.21	1.18
Low	0.183	0.149	0.033	18	5.44	2.28	0.41	2.98	1.25
SE	0.011	0.009	0.010	4.9	0.72	0.44	0.04	0.35	0.20
CV, %	5.5	5.6	35.9	33.1	12.2	21.4	11.0	11.0	18.6
				1984					
Spacing									
0.38	0.266	0.187**	0.078	29*	9.52*	3.33	0.35**	3.58**	1.25
0.76	0.265	0.175	0.089	34	8.13	3.29	0.40	3.08	1.25
Population									
High	0.263	0.184	0.078	30	9.41	3.39	0.36	3.56	1.28
Medium	0.269	0.183	0.085	31	8.45	3.20	0.38	3.16	1.19
Low	0.264	0.177	0.087	33	8.61	3.34	0.39	3.26	1.27
SE	0.013	0.008	0.012	35	1.05	0.50	0.02	0.33	0.18
CV, %	5.0	4.5	14.8	11.2	11.9	15.0	6.3	10.0	14.5

*,** Significant at 0.05 and 0.01, respectively.

† Significant ($P < 0.05$) spacing × population interactions in 1983 in harvest index and grain WUE. No significant spacing × population interactions in 1984.

wide row or low population studies. In these cases, higher water use was attributed either to higher net radiation or to advection from the dry soil surface between the wide rows. Yao and Shaw (1964a, b) showed higher water use for corn in wide rows and attributed it to higher net radiation over wide rows in Ames, IA. Steiner (1987) showed net radiation to be higher over wide than over narrow-row sorghum, but had earlier reported no significant water-use difference (Steiner, 1986).

The effect of row orientation on water use has been reported for corn and sorghum. Yao and Shaw (1964a, b) showed significantly greater water use for corn in E-W oriented rows than for N-S in Ames, IA. Steiner (1986), in a sorghum study at Bushland, TX, observed slightly more water use in E-W rows in both years, though not significant at the 5% level of probability. The E-W orientation had 14% higher net radiation than the N-S orientation in 1984, a moderate year. One should keep in mind that the prevailing wind in both Iowa and Texas is largely westerly. Therefore, advection may have contributed to the results. There is apparently a difference, therefore, that can be distinguished under certain circumstances of climate, soil, and crop, but it has not yet been studied sufficiently to describe.

Narrow rows have failed to increase yields where seasonal water supply was limited and when interaction of climate with days to maturity and population was not favorable for the season (Alessi and Power, 1974; Mitchell, 1970). Some have concluded sorghum is more suited to these dryland situations because of reduced water loss rates (Brown and Schroder, 1959; Olson, 1971). For the same reason, lower populations are frequently more successful under dryland conditions since they conserve early season water (Alessi and Power, 1965; Bond et al., 1964; Termunde et al., 1963).

Nutrient Use

The impact of planting geometries on nutrient relations for corn and grain sorghum has been studied, but reported less frequently than effects on water use, light interception, and plant growth. Presumably, this has occurred because most studies showed that row width caused no significant differences in grain and/or silage protein content (Bryant and Blaser, 1968; Cummins and Dobson, 1973; Karlen and Camp, 1985; Karlen et al., 1985, 1987; Lutz and Jones, 1969; Rhoads and Stanley, Jr. 1978; Stickler, 1964; Stickler and Laude, 1960). However, a second reason may be that many of the studies were conducted using high-fertilization rates or at soil-fertility levels that were considered nonlimiting (Lutz et al., 1971; Nunez and Kamprath, 1969; Rutger and Crowder, 1967; Stanley and Rhoads, 1971, 1974; Stivers et al., 1971). Karlen and Camp (1985) saw no differences in corn leaf concentrations at anthesis of N, P, K, Ca, Mg, S, B, Cu, Mn, or Zn among row-space variables.

Determining optimum planting geometries for enhanced nutrient uptake and utilization may become more important in the future. Public concern regarding the declining quality of groundwater resources is increasing (CAST, 1985; Keeney, 1986). Many significant increases in groundwater NO_3^- are the direct result of poor N fertilizer-use efficiencies. By combining precision fertilizer placement techniques, such as the spoke-injector technique (Baker et al., 1985), with improved planting geometries, fertilizer recovery may significantly increase and potential for groundwater contamination decrease (Touchton and Sims, 1987). Twin-row planting may have nutrient recovery advantages that were not apparent in the initial yield and light interception studies. In southeastern Coastal Plain soils, in-row subsoiling often results in a concentration of plant roots directly beneath the row because of a more favorable physical rooting environment (Campbell et al., 1984). In these soils, precision placement with a spoke-injector applicator may significantly increase the efficiency of N recovery by row crops.

Optimizing planting geometries may also become more important for the development of profitable agriculture production systems. By determining optimum plant spacing and populations for individual soil types or mapping units, fertilizer, herbicide, and irrigation applications can be managed more efficiently. The use of controlled traffic patterns and/or tramlines will also increase the importance of planting geometries. Using alternative planting geometries to enhance nutrient utilization of corn and sorghum may also become important when leguminous plants are grown in association to provide N, reduce soil erosion, recover residual fertilizer N, and supply subsoil water and nutrients to the primary crop (Blevins, 1987; Power, 1987).

Denser canopies are beneficial for soil erosion control. Greater uniformity and density of plant cover provides rainfall interception over a greater fraction of the soil surface. This reduces the velocity and hence kinetic energy of the rain drops, which in turn reduces the amount of soil dislodged at the soil surface. These effects prevent filling of macropores with soil debris from runoff and promote higher infiltration rates than under more open canopies (Mitchell, 1970; Pendleton, 1966) resulting in more efficient use of sprinkler irrigation or rainfall. A benefit of narrow rows in reducing furrow erosion has also been seen (Sojka and Brown, 1987). This resulted from shorter set times, energy dissipation by foliage intrusion, furrow lining by brace roots, fibrous root binding of aggregates at the furrow-water interface, and an increase in the infiltration of applied water.

OTHER CROPS

The topic of planting geometry has been addressed in a number of other crops. It would be beyond the scope of this chapter to review all of them in detail, but several unifying concepts and innovations are mentioned here briefly.

The twin-row concept has been adopted for use in wheat (*Triticum aestivum* L.) production in the Pacific Northwest. In this application, twin rows are coupled with banding of fertilizer between the rows (Veseth, 1987). This placement efficiently supplies nutrients to the intra-twin row soil, but "hides" nutrients from weeds in the inter-twin row area beyond the flanking pair of wheat rows. Early indications are that this practice can be used successfully with conservation tillage to limit weed growth and thereby increase the efficiency of nutrient and water use by the crop.

Row orientation has been studied in wheat with some interesting results (Erickson et al., 1979; Kirkham, 1980, 1982; Santhirasegaram and Black, 1968). Kirkham (1980) found that leaf orientation in wheat is cultivar dependent and that row orientation affected growth and light interception of winter wheat (Kirkham, 1982). During the winter, N-S rows had wide, short leaves and E-W rows had long narrow leaves. The N-S rows received more light than E-W rows, but less light, rather than more light, was associated with greater grain production. Santhirasegaram and Black (1968) determined that maximum light absorption occurred at 1200 h for E-W rows and at morning and evening for N-S rows.

Planting geometries have become a highly researched topic in peanut (*Arachis hypogaea* L.) production (Alexander, 1970; Chin Choy et al., 1977; Cook, 1980; Hauser and Buchanan, 1981; Mozingo, 1984; Mozingo and Coffelt, 1984; Schubert et al., 1983; Shelton, 1978; Stone et al., 1985). In one study, there were no differences in quality or yield for irrigated treatments for a variety of planting patterns including solid planting, skip-row planting, and twin-row planting (Schubert et al., 1983). Mozingo and Coffelt (1984) reported numerically but not statistically higher yields for twin-row planting patterns over single rows. Stone et al. (1985) found that stomatal diffusive resistance was higher in response to greater water stress in narrow-row peanuts. This suggests that peanut, like soybean, depletes more soil water in narrow rows.

Planting geometry effects on rice (*Oryza sativa* L.) yields and cultural relations were reviewed by Chandler (1969). His summary stated that newer stiff-strawed, low to medium tillering, shorter varieties performed best with close spacing at all N levels. The taller leafy tropical varieties would yield better at somewhat closer spacings if N and light were abundant. Jones and Snyder (1987a, b) confirmed this, stating that narrow-row spacings increased grain yields for both tall and semidwarf plants when drill seeded and did not affect yields of the subsequent ratooned crop.

Halterlein (1983) reviewed results for spacing of edible bean (*Phaseolus vulgaris* L.) and found that nearly all previous studies showed substantial yield benefit to increased LAI and narrow rows. He cautioned, however, that adoption of the newer configurations might be difficult due to the need to accommodate mechanical harvesters designed for wide rows.

Planting geometry of cotton was studied and reviewed briefly by Bilbro (1981) and Mohamad et al. (1982). Their work and the work they reviewed indicate that cotton is particularly sensitive to cultural practice interactions (e.g., irrigation and fertility) with season and choice of plant ideotype in determining the success of denser planting patterns. It appears that shorter statured, semideterminate, early maturing cultivars are most responsive to narrow rows and/or high populations.

CONCLUSION

For a wide range of species it appears that a greater yield advantage may exist for dense, uniformly spaced canopies than for more open canopies, provided early weed management, water availability, and fertility are adequate. This advantage is more pronounced for plants with indeterminate growth habits. Manipulation of row-spacing configurations may be the most practical means of optimizing canopy geometry. Leguminous plants may require greater than expected mineral nutrient availability to maximize nodulation and N_2 fixation in these denser, management-intensive cropping systems. Row spacing and row orientation may affect canopy light quality and soil temperatures sufficiently to alter rhizobial-host interactions and N_2 fixation of leguminous plants. The interaction of planting geometry and the nutrient relations of the soil-plant system are not well documented. Due to earlier LAI increases, early season (and often total season) ET are frequently higher in denser canopies. Wider rows of some crops may have higher ET due to sensible heat transfer from dry inter-row soil surface, although there are conflicting data on this point. Plant physiological indices of water stress frequently indicate greater stress in narrow rows or denser canopies, particularly under nonirrigated conditions, probably due to greater early season water depletion. Soil-physical and equipment-engineering limitations may impose the ultimate practical and financial barriers to adoption of high-intensity, dense-canopy management systems.

REFERENCES

Aldrich, S.R., W.D. Scott, and E.R. Leng. 1976. Modern corn production. 2nd ed. A&L Publ., Champaign, IL.

Alessi, J., and J.F. Power. 1965. Influence of moisture plant population, and nitrogen on dryland corn in the Northern Plains. Agron. J. 56:611–612.

----, and ----. 1974. Effects of plant population, row spacing and relative maturity on dryland corn in the Northern Plains. Agron. J. 66:316–319.

----, and ----. 1982. Effects of plant and row spacing on dryland soybean yield and water-use efficiency. Agron. J. 74:851–854.

Alexander, M.W. 1970. The relationship of growth habit and row pattern on yield and market grade of three Virginia peanuts. J. Am. Peanut Res. Educ. Assoc. 2:134 (Abstr.).

Andrews, R.H., and J.W. Peek. 1971. Influence of cultural practice and field environment on consistency of corn yields in northern areas. Agron. J. 63:628–633.

Aubertin, G.M., and D.B. Peters. 1961. Net radiation determinations in a corn field. Agron. J. 53:269–272.

Baker, J.L., T.S. Colvin, S.J. Marley, and D. Dawelbeit. 1985. Improve fertilizer management with a point-injector applicator. Paper 85-1516. ASAE, St. Joseph, MI.

Balatti, P.A., and E.R. Montaldi. 1983. Efecto del fotoperiodo sobre la nodulacion y fijacion de nitrogeno en plantas de soja (*Glycine max* L. Merr.). p. 23–29. *In* Revista de la Facultad de Agronomia, 3ª epoca, Tomo LVII, entrega 1–2. Univ. de La Plata, Argentina.

----, and ----. 1986. Effects of red and red/far-red lights on nodulation and nitrogen fixation in soybean (*Glycine max* L. Merr.). Plant Soil 92:427–430.

Basnet, B., E.L. Mader, and C.D. Nickell. 1974. Influence of between and within-row spacing on agronomic characteristics of irrigated soybeans. Agron. J. 66:657–659.

Beatty, K.D., I.L. Eldridge, and A.M. Simpson, Jr. 1982. Soybean response to different planting patterns and dates. Agron. J. 74:859–862.

Beaver, J.S., and R.R. Johnson. 1981. Response of determinate and indeterminate soybeans to varying cultural practices in the Northern U.S.A. Agron. J. 73:833–838.

Bello, A.B., W.A. Ceron-Diaz, C.D. Nickell, E.O. El Sherif, and L.C. Davis. 1980. Influence of cultivar, between-row spacing, and plant population of fixation. Crop Sci. 20:751–755.

Bennie, A.T.P., W.K. Mason, and H.M. Taylor. 1982. Responses of soybeans to two row spacings and two soil water levels: III. Concentration, accumulation and translocation of 12 elements. Field Crops Res. 5:31–43.

Bilbro, J.D. 1981. Spacial responses of contrasting cotton cultivars grown under semiarid conditions. Agron. J. 73:271–277.

Blad, B.L. 1983. Atmospheric demand for water. p. 1–44. *In* I.D. Teare and M.M. Peet (ed.) Crop water relations. John Wiley and Sons, New York.

Blevins, D.G. 1987. Future developments in plant nutrition research. p. 445–458. *In* L.L. Boersma (chmn.) Future developments in soil science research. SSSA, Madison, WI.

Bohm, W. 1977. Development of soybean root systems as affected by plant spacing. Z. Pflanzenbau 144:103–112.

Bond, J.J., T.J. Army, and O.R. Lehman. 1964. Row spacing, plant populations, and moisture supply as factors in dryland grain sorghum production. Agron. J. 56:3–6.

Boquet, D.J., K.L. Koonce, and D.M. Walker. 1982. Selected determinate soybean cultivar yield responses to row spacings and planting dates. Agron. J. 74:136–138.

Brown, P.L., and W.D. Schroder. 1959. Grain yields, evapotranspiration, and water use efficiency of grain sorghum under different cultural practices. Agron. J. 51:339–343.

Brown, R.H., E.R. Beatty, W.J. Ethredge, and D.D. Hayes. 1970. Influence of row width and plant population on yield of two varieties of corn (*Zea mays* L.). Agron. J. 62:767–770.

Bryant, A.A., R.C. Eckhardt, G.F. Sprague. 1940. Spacing experiments with corn. J. Am. Soc. Agron. 32:707–714.

Bryant, H.T., and R.E. Blaser. 1968. Plant constituents of an early and late corn hybrid as affected by row spacing and plant population. Agron. J. 60:557–559.

Burnside, O.C., and W.L. Colville. 1964. Soybean and weed yields as affected by irrigation, row spacing, tillage, and amiben. Weeds 12:109–112.

----, G.A. Wicks, and C.R. Fenster. 1964. Influence of tillage, row spacing, and atrazine on sorghum and weed yields from non-irrigated sorghum across Nebraska. Weeds 12:211–215.

Camp, C.R., D.L. Karlen, and J.R. Lambert. 1985. Irrigation scheduling and row configuration for corn in the southeastern coastal plain. Trans. ASAE 28:1159–1165.

Campbell, R.B., R.E. Sojka, and D.L. Karlen. 1984. Conservation tillage for soybean in the U.S. southeastern coastal plain. Soil Tillage Res. 4:531–541.

Camper, H.M., and T.J. Smith. 1958. The effect of date of planting, rate of planting, and width of row on two soybean varieties. Virginia Agric. Exp. Stn. Rep. 21.

Cardwell, V.B. 1982. Fifty years of Minnesota corn production: Sources of yield increase. Agron. J. 74:984–990.

Carter, T.E., Jr., and H.R. Boerma. 1979. Implications of genotype × planting date and row-spacing interactions in double-cropped soybean cultivar development. Crop Sci. 19:607–610.

CAST. 1985. Agricultural and groundwater quality. Report 103. CAST, Ames, IA.

Caviness, C.E. 1961. Effects of skips in soybean rows. Ark. Farm Res. 10:12.

----. 1966. Spacing studies with Lee soybeans. Arkansas Agric. Exp. Stn. Bull. 713.

----, and P.E. Smith. 1959. Effects of different dates and rates of planting soybeans. Arkansas Agric. Exp. Stn. Rep. Ser. 88.

Chan, L.M., R.R. Johnson, C.M. Brown. 1980. Relay intercropping soybeans into winter wheat and spring oats. Agron. J. 72:35–39.

Chandler, R.F., Jr. 1969. Plant morphology and stand geometry in relation to nitrogen. *In* J.D. Eastin et al. (ed.) Physiological aspects of crop yield. ASA and CSSA, Madison, WI.

Chin Choy, E.W., and E.T. Kanemasu. 1974. Energy balance comparison of wide and narrow row spacings in sorghum. Agron. J. 66:98–99.

----, J.F. Stone, and J.E. Garton. 1977. Row spacing and direction effects on water uptake characteristics of peanuts. Soil Sci. Soc. Am. J. 41:428–432.

Colville, W.L. 1966. Plant populations and row spacing. p. 55–62. *In* Annu. Hybrid Corn Ind. Res. Conf. Proc. 21.

----. 1968. Influence of plant spacing and population on aspects of the microclimate within corn ecosystems. Agron. J. 60:65–67.

----, and J.D. Furrer. 1964. Narrow spacings increase yields p. 7–9. *In* Nebraska Exp. Stn. Quart. W.

Cook, J. 1980. Skip-row peanut yields were good but not good enough. The Peanut Farmer 16:22–23.

Cooper, R.L. 1971. Influence of soybean production practices on lodging and seed yield in highly productive environments. Agron. J. 63:490–493.

----. 1977. Response of soybean cultivars to narrow rows and planting rates under weed-free conditions. Agron. J. 69:89–92.

----, and D.L. Jeffers. 1984. Use of nitrogen stress to demonstrate the effect of yield limiting factors on the yield response of soybean to narrow row systems. Agron. J. 76:257–259.

Costa, J.A., E.S. Oplinger, and J.W. Pendleton. 1980. Response of soybean cultivars to planting patterns. Agron. J. 72:153–156.

Cummins, D.G., and J.W. Dobson. 1973. Corn for silage as influenced by hybrid maturity, row spacing, plant population, and climate. Agron. J. 65:240–243.

deMooy, C.J. and J. Pesek. 1966. Nodulation responses of soybeans to added phosphorous, potassium, and calcium salts. Agron. J. 58:275–280.

Denmead, O.T., L.J. Fritschen, and R.H. Shaw. 1962. Spacial distribution of net radiation in a corn field. Agron. J. 53:505–510.

Donald, C.M. 1958. Interaction of competition for light and for nutrients. Austr. J. Agric. Res. 9:421–435.

Donovan, L.S. F. Dimmock, and R.B. Carson. 1963. Some effects of planting pattern on yield, percent oil and percent protein in Mandarin (Ottawa) soybeans. Can. J. Plant Sci. 43:131–140.

Doss, B.D., and D.L. Thurlow. 1974. Irrigation, row width, and plant population in relation to growth characteristics of two soybean varieties. Agron. J. 66:620–623.

Dougherty, C.T. 1969. The influence of planting date, row-spacing, and herbicides on the yield of soybeans in Canterbury. N. Z. J. Agric. Res. 12:703–726.

Downey, L.A. 1971. Plant density-yield relations in maize. J. Aust. Inst. Agric. Sci. 37:138–146.

Duncan, W.G. 1958. The relationship between corn population and yield. Agron. J. 50:82–84.

----. 1969. Cultural manipulation for higher yields. *In* J.D. Eastin et al. (ed.) Physiological aspects of crop yield. ASA and CSSA, Madison, WI.

----. 1972. Plant spacing, density, orientation, and light relationships as related to different corn genotypes. *In* Proc. 27 Annu. Corn Sorghum Res. Conf. ASTA, Washington.

Dungan, G.H., A.L. Lang, and J.W. Pendleton. 1958. Corn plant population in relation to soil productivity. Adv. Agron. 10:435–473.

Erickson, P.I., M.B. Kirkham, and J.F. Stone. 1979. Growth, water relations, and yield of wheat planted in four row directions. Soil Sci. Soc. Am. J. 43:570–574.

Fellers, C.R. 1918. The effect of inoculation, fertilizer, and certain minerals on the yield, composition, and nodule formation of soybeans. Soil Sci. 6:81–129.

Felton, W.L. 1976. The influence of row-spacing and plant population on the effect of weed competition in soybean. Aust. J. Exp. Agric. Anim. Husb. 16:926–931.

Fontes, L.A.N., and A.J. Ohlrogge. 1972. Influence of seed size and population on yield and other characteristics of soybean (*Glycine max* (L.) Merr.). Agron. J. 64:833–836.

Frans, R.E. 1959. Effect on soybean yields of herbicide and narrow row width combinations. Arkansas Agric. Exp. Stn. Rep. Ser. 4.

Gebhardt, M.R., and H.C. Minor. 1983. Soybean production systems for claypan soils. Agron. J. 75:532–537.

Grimes, D.W., and J.T. Musick. 1959. How plant spacing, fertility, and irrigation affect grain sorghum production in southwestern Kansas. Kansas Garden City Branch, Agric. Exp. Stn. Bull. 414.

Halterlein, A.J. 1983. Bean. p. 157–186. *In* I.D. Teare and M.M. Peet (ed.) Crop water relations. John Wiley and Sons, New York.

Hanks, R.J., L.H. Allen, and H.R. Gardner. 1971. Advection and evapotranspiration of wide-row sorghum in the Central Great Plains. Agron. J. 63:520–527.

Hartwig, E.E. 1954. Factors affecting time of planting soybeans in the southern states. USDA Circ. 943.

----. 1957. Row width and rates of planting in the southern states. Soybean Dig. 17:13–14.

Hauser, E.W., and G.A. Buchanan. 1981. Influence of row spacing, seeding rate, and herbicide systems on the competitiveness and yield of peanuts. Peanut Sci. 8:74–81.

Healey, R.G. 1985. Competition for land in the American South. The Conservation Foundation, Washington, DC.

Heltz, G.E., and A.L. Whiting. 1928. Effect of fertilizer treatment on the formation of nodules on the soybean. J. Am. Soc. Agron. 20:975–981.

Hicks, D.R., J.W. Pendleton, R.L. Bernard, and T.J. Johnston. 1969. Response of soybean plant type to planting patterns. Agron. J. 61:290–293.

Hinkle, D.A., and J.D. Garret. 1961. Corn fertilizer and spacing experiments. Arkansas Agric. Exp. Stn. Bull. 635.

Hinson, K., and W.D. Hanson. 1962. Competition studies in soybeans. Crop Sci. 2:117-123.

Hoff, D.J., and H.J. Mederski. 1960. Effect of equidistant corn plant spacing on yield. Agron. J. 45:490-493.

Hoggard, A.L., J.G. Shannon, and D.R. Johnson. 1978. Effect of plant population on yield and height characteristics in determinate soybeans. Agron. J. 70:1070-1072.

Holmes, M.G. 1981. Spectral distribution of radiation within plant canopies. p. 147-158. *In* H. Smith (ed.) Plants and the daylight spectrum. Academic Press, New York.

Howe, O.W., III, and L.R. Oliver. 1987. Influence of soybean (*Glycine max*) row-spacing on pitted morning glory (*Ipomoea lacunosa*) interference. Weed Sci. 35:185-193.

Hozumi, K., H. Koyami, and T. Kira. 1955. Interspecific competition among higher plants. IV. A preliminary account of the interaction between adjacent individuals, Ser. D., Vol. 6. Jpn. Inst. Polytechnics, Osaka City Univ., Japan.

Hunt, P.G., R.E. Sojka, T.A. Matheny, and A.G. Wollum II. 1985. Soybean response to *Rhizobium japonicum* strain, row orientation, and irrigation. Agron. J. 77:720-725.

Johnson, B.J., and H.B. Harris. 1967. Influence of plant population on yield and other characteristics of soybeans. Agron. J. 59:447-449.

Jones, D.B., and G.H. Snyder. 1987a. Seeding rate and row spacing effects on yield and yield components of drill-seeded rice. Agron. J. 79:623-626.

----, and ----. 1987b. Seeding rate and row spacing effects on yield and yield components of ratoon rice. Agron. J. 79:627-629.

Jordan, H.V., K.D. Laird, and D.D. Ferguson. 1950. Growth rates and nutrient uptake by corn in a fertilizer-spacing experiment. Agron. J. 42:261-268.

Kanemasu, E.T., and G.F. Arkin. 1974. Radiant energy and light environment of crops. Agric. Meteorol. 14:211-225.

Karlen, D.L., and C.R. Camp. 1985. Row spacing, plant population, and water management effects on corn in the Atlantic Coastal Plain. Agron. J. 77:393-398.

----, ----, and J.P. Zublena. 1985. Plant density, distribution, and fertilizer effects on yield and quality of irrigated corn silage. Commun. Soil Sci. Plant Anal. 16:55-70.

----, and M.J. Kasperbauer. 1988. Row orientation and configuration effects on spectral patterns and corn yield components. (Unpublished data.)

----, ----, and J.P. Zublena. 1987. Row-spacing effects on corn in the Southeastern U.S. Appl. Agric. Res. 2:65-73.

----, and R.E. Sojka. 1985. Hybrid irrigation effects on conservation tillage corn in the Coastal Plain. Agron. J. 77:561-567.

Kasperbauer, M.J., P.G. Hunt, and R.E. Sojka. 1984. Photosynthate partitioning and nodule formation in soybean plants that received red/far-red light at the end of the photosynthetic period. Physiol. Plant 61:549-554.

Keeney, D.R. 1986. Sources of nitrate to groundwater. CRC Crit. Rev. in Environ. Control 16:257-304.

Kirkham, M.B. 1980. Leaf orientation. Annu. Wheat News 26:130.

----. 1982. Orientation of leaves of winter wheat planted in north-south or east-west rows. Agron. J. 74:893-898.

Knipmeyer, J.W., R.H. Hageman, E.B. Earley, and R.D. Seif. 1962. Effect of light intensity on certain metabolites of the corn plant (*Zea mays* L.). Crop Sci. 2:1-5.

Kohnke, H., and S.R. Miles. 1951. Rates and patterns of seeding corn on high fertility land. Agron. J. 43:488-493.

Kust, C.A., and R.R. Smith. 1969. Interaction of linuron and row spacing for control of yellow foxtail and barnyard-grass in soybeans. Weed Sci. 17:489-491.

Lang, A.L., J.W. Pendleton, and G.H. Dungan. 1956. Influence of population and nitrogen levels on yield and protein and oil contents of nine corn hybrids. Agron. J. 48:284-289.

Larson, W.E., and J.J. Hanway. 1977. Corn production. *In* G.F. Sprague (ed.) Corn and corn improvement. Agronomy 18:625-669.

Laude, H.H., A.W. Pauli, and G.O. Throneberry. 1955. Crops and soils field day report. Kansas Agric. Exp. Stn. Circ. 323:21-23.

Leffel, R.C., and G.W. Barber. 1961. Row widths and seeding rates in soybeans. Maryland Agric. Exp. Stn. Bull. 470.

Lehman, W.F., and J.W. Lambert. 1960. Effects of spacing of soybean plants between and within rows on yield and its components. Agron. J. 52:84-86.

Lie, T.A. 1969. Non-photosynthetic effects of red/far-red light on root-nodule formation by leguminous plants. Plant Soil 30:391-404.

Loomis, R.S., and W.A. Williams. 1969. Productivity and the morphology of crop stands: Patterns with leaves. p. 27-52. *In* J.D. Easten et al. (ed.) Physiological aspects of crop yield. ASA and CSSA, Madison, WI.

Ludecke, H. 1941. Die Bedeutung der Phosphorsaure fur das Wachstum der Sojabohne und die Tatigkeit iher Knollchenbaktreien. Phosphorsaure 10:196-204.

Lueschen, W.E., and D.R. Hicks. 1977. Influence of plant population on field performance of three soybean cultivars. Agron. J. 69:390-393.

Lutz, J.A., Jr., H.M. Camper, and G.D. Jones. 1971. Row spacing and population effects on corn yields. Agron. J. 63:12-14.

----, and C.D. Jones. 1969. Effect of corn hybrids, row spacing, and plant population on the yield of corn silage. Agron. J. 61:942-945.

Malik, N.S.A., H.E. Calvert, M. Pence, and W.D. Bauer. 1982. Regulation of nodulation on soybean roots. Annu. Rep. Charles F. Kettering Res. Lab., Yellow Springs, OH.

Mason, W.K., A.T.P. Bennie, H.R. Rowse, T.C. Kaspar, and H.M. Taylor. 1982. Response of soybeans to two row spacings and two soil water levels. II. Water use, root growth, and plant water status. Field Crops Res. 5:15-29.

----, H.M. Taylor, A.T.P. Bennie, H.R. Rowse, D.C. Reicosky, Y.S. Jung, A.A. Righes, R.L. Yang, T.C. Kaspar, and J.A. Stone. 1980. Soybean row spacing and soil water supply: Their effect on growth, development, water relations, and mineral uptake. p. 1-59. *In* Adv. Agric. Technol. Publ. SEA-NC-5. AR, North Central region, USDA-SEA, Peoria, IL.

McCauley, G.N., J.F. Stone, and E.W. Chin Choy. 1978. Evapotranspiration reduction by field geometry effects in peanuts and grain sorghum. Agric. Meteorol. 19:295-304.

Mitchell, R.L. 1970. Crop growth and culture. The Iowa State University Press, Ames.

Mock, J.J., and L.C. Heghin. 1976. Performance of maize hybrids grown in conventional row and randomly distributed planting patterns. Agron. J. 68:577-580.

Mohamad, K.B., W.P. Sappenfield, and J.M. Poehlman. 1982. Cotton cultivar response to plant populations in a short season, narrow-row cultivar system. Agron. J. 74:619-625.

Mooers, C.A. 1910. Stand and soil fertility as factors in the testing of varieties of corn. Tennessee Agric. Exp. Stn. Bull. 89.

Morrow, G.E. 1890. Field experiments with corn. Illinois Agric. Exp. Stn. Bull. 13.

Mozingo, R.W. 1984. Skip-row planting and row pattern effects on Virginia-type peanut cultivars. Agron. J. 76:660-662.

----, and T.A. Coffelt. 1984. Row pattern and seeding rate effects on value of Virginia type peanut. Agron. J. 76:460-462.

Munevar, F., and A.G. Wollum II. 1981a. Effect of high root temperature and Rhizobium strain on nodulation, nitrogen fixation, and growth of soybeans. Soil Sci. Soc. Am. Proc. 45:1113-1120.

----, and ----. 1981b. Growth of *Rhizobium japonicum* strains at temperatures above 27 C. Appl. Environ. Microbiol. 42:272-276.

----, and ----. 1982. Response of soybean plants to high root temperatures as affected by plant cultivar and rhizobium strain. Agron. J. 74:138-142.

Nave, W.R., and R.L. Cooper. 1974. Effect of plant population and row width on soybean yield and harvesting loss. Trans. ASAE 17:801-804.

Nelson, M. 1931. Preliminary report on cultural and fertilizer experiments with rice in Arkansas. Arkansas Agric. Exp. Stn. Bull. 264.

Nunez, R., and E.J. Kamprath. 1969. Relationships between N response, plant population, and row width on growth and yield of corn. Agron. J. 61:279-282.

Olson, T.C. 1971. Yield and water use by different populations of dryland corn, grain sorghum, and forage sorghum in the western cornbelt. Agron. J. 63:104-106.

Painter, C.G., and R.W. Leamer. 1953. The effect of moisture, spacing, fertility, and their interrelationships on grain sorghum production. Agron. J. 45:261-264.

Parker, M.B., W.H. Marchant, and B.J. Mullinix, Jr. 1981. Date of planting and row-spacing effects on four soybean cultivars. Agron. J. 73:759-762.

Pendleton, J.W. 1966. Increasing water use efficiency by crop management. *In* W.H. Pierre et al. (ed.) Plant environment and efficient water use. ASA and SSSA, Madison, WI.

----, and D.B. Peters, and J.W. Peek. 1966. Role of reflected light in the corn ecosystem. Agron. J. 58:73-74.

Peters, D.B. 1965. Interrelations of row spacing, cultivations, and herbicides for weed control in soybeans. Weeds 13:285-289.

----, and L.C. Johnson. 1960. Soil moisture use by soybeans. Agron. J. 52:687-699.

----, and M.B. Russell. 1959. Relative water losses by evaporation and transpiration in field corn. Soil Sci. Soc. Am. Proc. 23:170-173.

Peters, E.J., M.R. Gebhardt, and J.F. Stritzke. 1965. Interrelations of row spacings, cultivations, and herbicides for weed control in soybeans. Weeds 13:285-289.

Plaut, Z., A. Blum, and I. Arnon. 1969. Effect of soil moisture regime and row spacing on grain sorghum production. Agron. J. 61:344-347.

Porter, K.B., M.E. Jensen, and W.H. Sletten. 1960. The effect of row spacing, fertilizer and planting rate on the yield and water use of irrigated grain sorghum. Agron. J. 52:431-434.

Poschenrieder, H., K. Sammet, and R. Fischer. 1940. Untersuchungen uber den Einflusz verschiedener Eranhrung mit Kali und Phosphorsaure auf die Aufbildung der Wurzelknollchen und die Tatigkeit der Knolchenbakterien bei der Sojabohne. Zentralbl. Bakteriol. II 102:388-395, 425-432.

Power, J.F. (ed.) 1987. The role of legumes in conservation tillage systems. *In* Proc. Nat. Conf., Univ. of Georgia, Athens. 27-29 April. Soil Conserv. Soc. Am., Ankeny, IA.

Prasad, K., R.C. Gautam, and N.K. Mohta. 1985a. Effect of planting patterns and weed control methods on growth characters, yield and yield attributes of ahar intercropped with soybean. Indian J. Agron. 30:429-433.

----, ----, and ----. 1985b. Studies on weed control in arhar and soybean as influenced by planting patterns, intercropping and weed control methods. Indian J. Agron. 30:434-439.

Probst, A.H. 1945. Influence of spacing on yield and other characteristics in soybeans. J. Am. Soc. Agron. 37:549-554.

Ramseur, E.L., V.L. Quisenberry, S.U. Wallace, and J.H. Palmer. 1984. Yield and yield components of 'Braxton' soybeans as influenced by irrigation and intra-row spacing. Agron. J. 76:442-446.

Reicosky, D.C., T.C. Kaspar, and H.M. Taylor. 1982a. Diurnal relationship between evapotranspiration and leaf water potential of field-grown soybeans. Agron. J. 74:667-673.

----, H.R. Rowse, W.K. Mason, and H.M. Taylor. 1982b. Effect of irrigation and row spacing on soybean water use. Agron. J. 74:958-964.

----, D.D. Warnes, and S.D. Evans. 1985. Soybean evapotranspiration, leaf water potential and foliage temperature as affected by row spacing and irrigation. Field Crops Res. 10:37-48.

Reynolds, E.B. 1926. The effect of spacing on the yield of cotton. Texas Agric. Exp. Stn. Bull. 340.

Rhoads, F.M., and R.L. Stanley, Jr. 1978. Effect of population and fertility on nutrient uptake and yield components of irrigated corn. Soil Crop Sci. Soc. Fla. Proc. 38:78-81.

Rutger, J.N., and L.V. Crowder. 1967. Effect of high plant density on silage and grain yields of six corn hybrids. Crop Sci. 7:182-184.

Ryder, G.J., and J.E. Beuerlein. 1979. Soybean production—a system approach. Crops Soils 31:9-11.

Safo-Kantanka, O., and N.C. Lawson. 1980. The effect of different row spacings and plant arrangements on soybeans. Can. J. Plant Sci. 60:227-231.

Santhirasegaram, K., and J.N. Black. 1968. The distribution of leaf area and light intensity within wheat crops differing in row direction, row-spacing and rate of sowing; a contribution to the study of undersowing pasture with cereals. J. Br. Grassl. Soc. 23:1-12.

Schubert, A.M., C.L. Pohler, and D.H. Smith. 1983. Skip-row peanuts in south central Texas. Texas Agric. Exp. Stn. Proj. Rep. 4058.

Sentz, J.C. 1965. Effect of row width on corn production. Minn. Farm Home Sci. 22:3-5.

Shaw, R.H., and C.R. Weber. 1967. Effects of canopy arrangements on light interception and yield of soybeans. Agron. J. 59:155-159.

Shelton, A. 1978. In far west Texas spanish peanuts yield 7640 lbs an acre on skip-row planting. The Peanut Farmer 14:8-10.

Shibles, R., I.C. Anderson, and A.H. Gibson. 1975. Soybeans. p. 151. *In* L.T. Evans (ed.) Crop physiology, some case histories. Cambridge Univ. Press, Cambridge, England.

Shibles, R.M., and C.R. Weber. 1966. Interception of solar radiation and dry matter production by various soybean planting patterns. Crop Sci. 6:5-59.

Shubeck, F.E., and H.G. Young. 1970. Equidistant corn planting. Crops Soils 22:12-14.

Smith, P.E. 1952. Soybean yields as affected by row widths. Ark. Farm Res. 1:4.

Sojka, R.E., and M.J. Brown. 1987. Furrow erosion as influenced by furrow spacing and canopy configuration. p. 247. *In* Agronomy abstracts. ASA, Madison, WI.

----, C.R. Camp, J.E. Parsons, and D.L. Karlen. 1984a. Measurement variability in soybean water status and soil-nutrient extraction in a row spacing study in the U.S. southeastern coastal plain. Commun. Soil Sci. Plant. Anal. 15:1111-1134.

----, G.W. Langdale, and D.L. Karlen. 1984b. Vegetative techniques for reducing water erosion of cropland in the southern United States. Adv. Agron. 37:155-181.

----, and J.E. Parsons. 1983. Soybean water status and canopy microclimate relationships at four row spacings. Agron. J. 75:961-968.

Stanley, R.L., Jr., and F.M. Rhoads. 1971. Response of corn grown at low soil moisture tension to row and drill spacings. Fla. Soil Crop Sci. Soc. 31:45-48.

----, and ----. 1974. Response of corn (*Zea mays* L.) to population and spacing with plow-layer soil water management. Fla. Soil Crop Sci. Soc. 34:127-130.

Steiner, J.L. 1986. Dryland grain sorghum water use, light interception, and growth responses to planting geometry. Agron. J. 78:720-726.

----. 1987. Radiation balance of dryland grain sorghum as affected by planting geometry. Agron. J. 79:259-265.

Stickler, F.C. 1964. Row width and plant population studies with corn. Agron. J. 56:438-441.

----, and H.H. Laude. 1960. Effect of row spacing and plant population on performance of corn, grain sorghum, and forage sorghum. Agron. J. 52:275-277.

Stivers, R.K., D.R. Griffith, and E.P. Christmas. 1971. Corn performance in relation to row spacings, populations, and hybrids on five soils in Indiana. Agron. J. 63:580-582.

----, and M.L. Swearingin. 1980. Soybean yield compensation with different populations and missing plant patterns. Agron. J. 72:98-102.

Stone, J.F., P.I. Erickson, and A.S. Abdul-Jabber. 1985. Stomatal closure behavior induced by row-spacing and evaporative demand in irrigated peanuts. Agron. J. 77:197-202.

Stringfield, G.H. 1962. Corn plant population as related to growth conditions and to genotype. Proc. Hybrid Corn Ind. Res. Conf. 17:61-68.

Tanner, C.B. 1957. Factors affecting evaporation from plants and soils. J. Soil Water Conserv. 12:221-227.

----, A.E. Peterson, and J.R. Love. 1960. Radiant energy exchange in a corn field. Agron. J. 52:373-379.

Taylor, H.M. 1980. Soybean growth and yield as affected by row spacing and by seasonal water supply. Agron. J. 72:543-547.

----, W.K. Mason, A.T.P. Bennie, and H.R. Rowse. 1982. Responses of soybeans to two row spacings and two soil water levels: I. An analysis of biomass accumulation, canopy development, solar radiation interception and components of seed yield. Field Crops Res. 5:1-14.

Termunde, D.E., D.B. Shank, and V.A. Dicks. 1963. Effects of population levels on yield and maturity of maize hybrids grown in the Northern Great Plains. Agron. J. 55:551-555.

Timmons, D.R., R.F. Holt, and J.T. Moraghan. 1966. Effect of corn population on yield, evapotranspiration, and water use efficiency in the northwest corn belt. Agron. J. 58:429-432.

----, ----, and R.L. Thompson. 1967. Effect of plant population and row spacing on evapotranspiration and water-use efficiency of soybeans. Agron. J. 59:262-265.

Touchton, J.T., and J.T. Sims. 1987. Tillage systems and nutrient management in the east and southeast. p. 225-234. *In* L.L. Boersma (chmn.) Future developments in soil science research. SSSA, Madison, WI.

Veseth, R. 1987. Paired-row versus single-row spacing. p. 1-5. *In* R. Veseth and D. Wysocki (ed.) Steep extension conservation farming update. Winter. Univ. of Idaho, Moscow; Oregon State Univ., Pendleton; and Washington State Univ., Pullman.

Waldren, R.P. 1983. Corn. p. 187-211. *In* I.D. Teare and M.M. Peet (ed.) Crop water realtions. John Wiley and Sons, New York.

Washko, J.B., and W.L. Kjelgaard. 1966. Double row corn planting can increase silage yields. Sci. Farmer 13:7.

Wax, L.M., W.R. Nave, and R.L. Cooper. 1977. Weed control in narrow and wide-row soybeans. Weed Sci. 25:73-78.

----, and J.W. Pendleton. 1968. Effect of row spacing on weed control in soybeans. Weed Sci. 16:462-465.

Weber, C.R., R.M. Shibles, and D.F. Byth. 1966. Effect of plant population and row-spacing on soybean development and production. Agron. J. 58:99-102.

Wiggams, R.G. 1939. The influence of space and arrangement on the production of soybean plants. J. Am. Soc. Agron. 31:314-321.

Wilcox, J.R. 1974. Response of three soybean strains to equidistant spacings. Agron. J. 66:409-412.

Williams, C., L. Mason, and B.E. Newman. 1970. Response of four soybean varieties to varying plant population density and row spacing. p. 135-143. *In* Louisiana Agric. Exp. Stn. Dep. Agron. Rep. Proj.

Wilson, J.K. 1917. Physiological studies of *Bacillus radiciola* of soybeans and of factors influencing nodule production. Cornell Agric. Exp. Stn. Bull. 386.

Woolley, D.G., N.P. Baracco, and W.A. Russell. 1962. Performance of four corn inbreds in single-cross hybrids as influenced by plant density and spacing patterns. Crop Sci. 2:441-444.

Yao, A.Y.M., and R.W. Shaw. 1964a. Effect of plant population and planting pattern of corn on water use and yield. Agron. J. 56:147-152.

----, and ----. 1964b. Effect of plant population and planting pattern of corn on the distribution of net radiation. Agron. J. 56:165-169.

Yoda, K., T. Kira, and K. Hozumi. 1957. Interspecific competition among higher plants. IX. Further analysis of the competitive interaction between adjacent individuals. Ser. D., Vol. 8. Jpn. Inst. Polytechnics, Osaka City Univ., Japan.

5 Role of Crop Residues—Improving Water Conservation and Use[1]

P. W. Unger
USDA-ARS
Bushland, Texas

G. W. Langdale
USDA-ARS
Watkinsville, Georgia

R. I. Papendick
USDA-ARS
Pullman, Washington

Water storage in soil is essential because most plants use water from the soil between precipitation or irrigation events (some plants store water in their tissues). Adequate water storage in soil is important not only in drier regions (arid, semiarid, and subhumid) but also in humid regions where short-term droughts at critical growth stages can greatly reduce crop yields. Because of their importance, water storage in soil and subsequent efficient use of that water for crop production have long been researched.

Early water conservation research generally involved the effects of tillage methods, row spacings, rotations, fallowing, etc. on responses of various crops to these practices. In those studies, the emphasis usually was on crop residue removal and/or incorporation (clean tillage). The value of surface residues for enhancing soil water storage generally was not recognized until the 1930s. Since then, numerous studies involving surface residues have been conducted. A complete review of all literature relative to residue effects on water conservation and use is beyond the scope of this chapter. Hence, we will give an overview of the early results, discuss results from present-day residue management systems, and discuss research needs and goals for further improving water conservation and use under surface residue conditions.

[1] Contribution from USDA-ARS, Conservation and Production Res. Lab., P.O. Drawer 10, Bushland, TX 79012; Southern Piedemont Conserv. Res. Ctr., Watkinsville, GA 30677; and Land Management and Water Conserv. Res., Pullman, WA 99164.

Cropping Strategies for Efficient Use of Water and Nitrogen, Special Publication no. 51.

OVERVIEW OF EARLY STUDIES WITH SURFACE RESIDUES

The value of a surface cover of plant materials for soil and water conservation purposes has long been recognized (Barnett, 1987; Bennett, 1939; Jenny, 1961; Lowdermilk, 1953). However, in early studies, interest in water conservation was related to reducing runoff so that soil losses would be reduced and not generally for the purpose of increasing soil water storage and subsequent use of that water for crop production.

One of the earliest reports concerning surface residue effects on increasing water storage was that of Hallsted and Mathews (1936), who said that "trash and crop residues on the surface check runoff and allow more of the water to be absorbed by the soil." Studies involving various surface residue treatments soon followed. Duley and Kelly (1939) measured simulated rainfall infiltration into six soils ranging in surface textures from sandy loam to clay loam and subsoil textures from sand to clay. The soils were bare (cultivated) or covered with wheat (*Triticum aestivum* L.) straw at a 5.6-Mg ha^{-1} rate. Total infiltration and final infiltration rates averaged about five and three times greater, respectively, with straw cover than with bare surface treatments. These differences were attributed to a thin dense layer that formed on the surface of bare soil but not on a straw-covered soil. Duley and Kelly (1939) concluded that surface conditions affected infiltration more than soil type, slope, previous water content, and rainfall intensity. Water contents were not reported, but the results suggested that more rainwater would be stored in straw-covered than in bare soils, provided, of course, that there was potential to store the additional water.

The study by Duley and Russel (1939) involved straw management (surface vs. incorporated) and tillage treatments (Table 5-1). Water storage with the surface straw treatment was about 2.7 and 2.0 times greater than for the plowed (no straw) and basin listed (diked or dammed furrow) treatments, respectively. Basin listing prevented runoff, yet water storage with it was less than with the surface straw treatment. This indicated that preventing runoff was not the total solution to water conservation in the Great Plains and that evaporation suppression by surface residues was a major component in soil water conservation. In other studies by Duley and Russel (1939), water storage increased from 50 to 80 mm as a result of surface-applied straw as compared with incorporated straw treatments.

The contribution of surface residues for conserving water by reducing runoff and evaporation was illustrated by Russel (1939). In a field study (Table 5-2), water storage was greatest and evaporation was least where 17.9 Mg ha^{-1} of straw was on the surface. Evaporation was highest with basin listing. These results substantiated earlier results which showed that favorable infiltration and reduced evaporation were important factors for improving soil water conservation.

At the time of the above studies, large-scale field operations with large amounts of surface residues were not practical because suitable equipment for controlling weeds, maintaining crop residues on the surface, and planting crops in surface residues had not yet been developed. Also, chemicals

Table 5-1. Effect of straw and tillage treatments on soil water storage, 23 April to 9 Sept. 1938 at Lincoln, NE (Duley and Russel, 1939).

Treatment	Precipitation stored†		Depth of water penetration
	mm	%	m
Straw, 4.5 Mg ha^{-1}, on surface	247	54.3	1.8
Straw, 4.5 Mg ha^{-1}, disked in	176	38.7	1.5
Straw, 4.5 Mg ha^{-1}, plowed in	155	34.1	1.5
No straw, disked	89	19.6	1.2
No straw, plowed	94	20.7	1.2
Decayed straw, 2 Mg ha^{-1}, plowed in	79	17.4	1.2
Basin listed	126	27.7	1.5

† Precipitation totaled 455 mm.

Table 5-2. Water storage, runoff, and evaporation from field plots at Lincoln, NE, 10 Apr. to 27 Sept. 1939 (Russel, 1939).

Treatment	Storage	Runoff	Evaporation	Evaporative loss†
	mm			%
Straw, 2.2 Mg ha^{-1}, normal subtillage	30	26	265	83
Straw, 4.5 Mg ha^{-1}, normal subtillage	29	10	282	88
Straw, 4.5 Mg ha^{-1}, extra loose subtillage	54	5	262	82
Straw, 9.0 Mg ha^{-1}, normal subtillage	87	trace	234	73
Straw, 17.9 Mg ha^{-1}, no tillage	139	0	182	57
Straw, 4.5 Mg ha^{-1}, disked in	27	28	266	83
No straw, disked	7	60	254	79
Contour basin listing	34	0	287	89

† Based on total precipitation, which was 321 mm for the period.

for controlling weeds were not yet available. However, those studies demonstrated the vast potential of surface residues for improving water conservation, and further studies involving surface residues soon were conducted at various locations.

Stubble Mulch Tillage

Stubble mulch "farming" or tillage (subtillage) is a crop production system involving surface residues that apparently was first used in the USA by a farmer in Georgia in the early 1930s for controlling water erosion in fields having slopes up to 17%. It greatly reduced runoff and erosion and quickly became a recommended soil conservation practice in the southeast USA (Middleton, 1952).

Stubble mulch tillage became an important wind erosion practice in the U.S. and Canadian Great Plains during the late 1930s and early 1940s. In that region, its value for reducing runoff and controlling water erosion also soon became apparent. McCalla and Army (1961) summarized the early research on stubble mulch farming and concluded that it was the most practical wind erosion control method and that it reduced water erosion when adequate residues were present. The limited effectiveness for controlling water erosion resulted primarily from the relatively low amounts of residue (generally <3.0 Mg ha^{-1}) produced by dryland (rainfed) crops in the Great Plains. Low residue production also contributed to the generally small increases in soil water storage and, subsequently, in relatively small increases in crop yields with stubble mulch tillage as compared with those obtained with clean tillage. These limitations, however, did not prevent stubble mulch tillage from becoming widely used for wheat production in semiarid regions of the USA, where it still is used extensively.

Stubble mulch tillage was evaluated also at some more humid locations where it was less adaptable than clean tillage for wheat because of problems in weed control, regulation of available N, and equipment operation under the conditions of wetter soils and higher amounts of surface residues (Zingg and Whitfield, 1957). Corn (*Zea mays* L.) grain yields were lower with stubble mulch tillage in New York (Free, 1953) but generally were little affected by tillage method (plowing, listing, subsurface tilling) in Iowa (Browning et al., 1945). Also, crop yields in the Southern Piedmont were little affected by tillage treatments (including stubble mulch tillage) (Beale et al., 1955; Peele et al., 1947).

Mulch and Tillage Studies

In addition to the early stubble mulch tillage studies, other early studies at various locations determined the effects of surface residues and tillage methods on runoff and soil losses. Soil water contents were not reported in all cases. However, higher infiltration and lower runoff with surface residues in the early studies (Borst and Woodburn, 1942; Browning et al., 1945; Chandler and Mason, 1942; Dawson, 1946; James, 1945; Kidder et al., 1943; Peele et al., 1947) as well as in more recent studies (Beale et al., 1955; Hays, 1961; Larson et al., 1960; Mannering and Meyer, 1961, 1963; Meyer and Mannering, 1961; Myhre and Sanford, 1972; Naffziger and Horner, 1958; Onstad, 1972; Taylor et al., 1964) indicated (or implied) that soil water contents were higher with than without the residue treatments, provided the soil had capacity to store the additional water. Crop yields were not reported for all the above studies, but yields generally were equal or higher with than without surface residues on well-drained soils and where seedbed preparation and planting were satisfactory. Also, surface residues greatly reduced soil losses, often more than they reduced runoff.

No-Tillage

No-tillage was used initially for pasture renovation, with close grazing or burning and heavy seeding rates to suppress competition between existing vegetation and the surface-sown forage species. Trampling by animals resulted in good seed-soil contact (Baeumer and Bakermans, 1973). In the 1940s, use of oils to control weeds in citrus (*Citrus* spp.) orchards was started in California (Johnston and Sullivan, 1949; Lombard, 1944). In the 1950s, Sprague (1952) used chemicals for weed control during pasture renovation. Also in the 1950s or early 1960s, no-tillage studies for field crops were initiated by Free et al. (1963), Lillard and Jones (1964), Moody et al. (1961), Shear and Moschler (1969), and Triplett et al. (1963) in the humid eastern USA and by Army et al. (1961), Baker et al. (1956), Barnes and Bohmont (1958), Barnes et al. (1955), Phillips (1954), Wiese and Army (1958, 1960), and Wiese et al. (1960) at several Great Plains locations.

In citrus orchards, the oil controlled weeds, reduced the irrigation requirement, and did not adversely affect fruit yields (Johnston and Sullivan, 1949; Lombard, 1944). For pasture renovation, an application of sodium trichloroacetate followed by two diskings was equally as effective as the normal practice of 10 to 12 diskings to kill the old sod and prepare the seedbed (Sprague, 1952).

No-Tillage—Humid Regions

The major goal of early no-tillage in the humid regions of the eastern USA was to control water erosion. Although less critical than for subhumid and semiarid regions, water conservation also was important because short-term droughts reduced yields sharply, especially on soils with a limited water storage capacity.

One of the earliest studies involving row-crop production by no-tillage methods in the USA was conducted in Virginia by Moody et al. (1961). They killed a grass-legume sod with atrazine (2-chloro-4-ethylamino-6-isopropyl-amino-1,3,5-triazine)[2] or a plastic cover about 5 weeks before hand-planting corn in holes made with a tube sampler. The sod was not disturbed. Soil water contents were greater with no-tillage throughout the growing season. Plant heights and stover yields also were greater with no-tillage, but grain yields with conventional and no-tillage treatments were similar. Yield increases with no-tillage as compared with conventional tillage were obtained by Lillard and Jones (1964) and Shear and Moschler (1969), also in Virginia. The yield increases were attributed to higher soil water contents. An additional benefit of no-tillage was effective erosion control due to the presence of surface residues.

Corn yield in New York (Free et al., 1963) was about 10% lower with no-tillage than with conventional tillage, but the difference was not statistically significant. Soil water contents were not reported, but no-tillage provided protection against erosion.

In the midwestern USA, runoff and erosion often are high because many soils have low infiltration rates and steep slopes. This sometimes causes low water storage followed by plant water stress and reduced yields. In early no-tillage studies in this region (Harrold et al., 1967, 1970; Van Doren and Triplett, 1969), water contents and corn yields often were higher and soil losses were lower with no-tillage than with conventional tillage.

[2] Mention of a trade name or product does not constitute a recommendation or endorsement for use by the USDA, nor does it imply registration under FIFRA as amended.

No-Tillage—Semiarid and Subhumid Regions

Major goals of no-tillage in the semiarid to subhumid Great Plains were improved surface residue maintenance for erosion control (mainly wind erosion) and improved water conservation. Improved wind erosion control was a definite advantage of early no-tillage, but effects on water conservation and crop yields were variable because herbicides sometimes did not control weeds during the interval (fallow period) between crops (Wiese and Staniforth, 1973). In addition, most studies involved dryland crops that did not produce sufficient residues to markedly enhance infiltration and suppress evaporation (Smika and Unger, 1986). For these reasons and because herbicide cost was more than tillage cost (Wiese, 1966), interest in no-tillage (chemical fallow) in the Great Plains during the 1960s was limited, and no-tillage research was conducted only at a few Great Plains locations during that decade.

Although no-tillage showed little promise in the Southern Great Plains, some progress occurred in the Central and Northern Great Plains when atrazine and propazine [2-chloro-4,6-bis(isopropylamino)-*s*-triazine] were used to control weeds in cropping systems involving grain sorghum [*Sorghum bicolor* (L.) Moench]. A common cropping sequence in the Great Plains that involves sorghum is the winter wheat (275 d)-fallow (335 d)-sorghum (150 d)-fallow (335 d) rotation, which results in two crops in 3 yr. Duration of the periods is approximate. Hereafter, the sequence is called WSF.

Phillips (1964) applied atrazine or propazine to separate plots after wheat harvest in a WSF sequence. In some cases, 2,4-D [(2,4-dichlorophenoxy) acetic acid] also was applied to control existing weeds. Atrazine gave excellent weed control. Propazine also gave good control, except where grassy weeds were present before it was applied. Several operations were needed to control weeds on tillage plots. Although soil water storage was similar, sorghum grain yields averaged 4.73 and 3.04 Mg ha^{-1} on herbicide-treated and tillage plots, respectively, undoubtedly because of better utilization of growing season precipitation with no-tillage.

Herbicide-Tillage Combinations

The study by Phillips (1964) showed that no-tillage was suitable for sorghum production in a WSF system when weed control was adequate. However, after a few years of no-tillage, the weed population shifted from mainly broadleaf species that were susceptible to atrazine to grassy species such as field sandbur [*Cenchrus longispinus* (Hackel) Fern.] that were resistant. The grassy weeds resulted in grain yields being similar for the two treatments. When atrazine-treated plots were tilled once at the time of atrazine application or at planting or cultivated once after sorghum establishment, no additional weed control measures were needed. Sorghum grain yields averaged 3.72 and 2.29 Mg ha^{-1} with the herbicide-tillage and tillage-only treatments, respectively (Phillips, 1969).

Although Phillips (1969) did not report soil water contents, higher grain yields with herbicide-tillage than with tillage-only treatments possibly resulted

Table 5-3. Effect of tillage and herbicide treatments on soil water contents at the end of the fallow period† and on wheat and sorghum yields in a wheat-fallow-sorghum rotation (Smika and Wicks, 1968).

Treatment from wheat harvest to sorghum planting		Treatment from sorghum harvest to wheat planting‡	Soil water storage§	Grain yields	
Fall	Spring			Wheat	Sorghum
			mm	Mg ha^{-1}	
Subtillage	Disk	Subtillage (5)	186b*	3.49a	4.08b
Subtillage	Atrazine	Subtillage (4)	213ab	3.76a	4.20b
Atrazine	Atrazine	Subtillage (4)	211ab	3.63a	4.58ab
Atrazine	Atrazine	Contact herbicide (4-6)	223a	3.49a	4.89a
Subtillage	Atrazine	Contact herbicide (4-6)	216ab	3.63a	5.02a

* Average values in a column followed by the same letter or letters are not significantly different at the 5% level (Duncan's multiple range test).
† Fallow duration of about 335 d.
‡ Values in parentheses denote number of operations.
§ Determined to a 3-m depth.

from higher water contents at planting and more effective use of growing-season precipitation. In studies by Smika and Wicks (1968) and Wicks and Smika (1973), water storage was less for tillage (plowing or disking) treatments that incorporated residues than for stubble mulch tillage, limited tillage, or herbicide-only treatments that maintained residues on the surface (Tables 5-3, 5-4, and 5-5). Water storage was highest for treatments involving herbicides because more residues were maintained on the surface and surface soil structure was not degraded by tillage, thus maintaining water infiltration at a higher rate. In general, crop yields increased with increases in water storage.

The foregoing examples illustrate that major increases in water storage and yields of subsequent crops are possible in semiarid to subhumid Great Plains by using practices that retain crop residues on the surface. The progressive improvement in water conservation and wheat grain yields at Akron,

Table 5-4. Effect of tillage and herbicide treatments on soil water contents at the end of the fallow period and on wheat yields in a 2-yr (wheat-fallow) rotation (Smika and Wicks, 1968).

Operations during fallow†		Soil water storage‡	Grain yield
Initial operation following wheat harvest	Subsequent operations		
		mm	Mg ha^{-1}
Plow	Subtillage (5)§	186c*	3.09b
Subtillage	Subtillage (5)	238b	3.36ab
Atrazine followed by subtillage	Subtillage (5)	272b	3.29ab
Atrazine	Subtillage (4)	275b	3.36ab
Atrazine	Contact herbicides (4-6)	325a	3.56a

* Average values in a column followed by the same letter or letters are not significantly different at the 5% level (Duncan's multiple range test).
† Fallow duration of about 425 d.
‡ Determined to a 3-m depth.
§ Values in parentheses denote number of operations.

Table 5-5. Effect of tillage and herbicide treatments on number of operations needed for weed control, residues maintained on the surface, soil water storage during fallow, and wheat yields in a 2-yr (wheat-fallow) rotation (Smika and Wicks, 1973).

Treatment	Operations during fallow†		Residues maintained§	Soil water storage¶	Grain yield
	Tillage‡	Herbicide application			
	—— no. yr^{-1} ——		%	mm	Mg ha^{-1}
Plow	8.5	0.0	0	146	2.69
Stubble mulch	8.7	0.0	21	203	2.88
Atrazine + stubble mulch	7.6	1.4	21	215	2.91
Atrazine + contact herbicide + stubble mulch	5.1	2.8	25	237	3.04
Atrazine + contact herbicide	0.0	6.0	46	274	3.17

† Fallow duration of about 425 d.
‡ The plow treatment included one moldboard plowing in the spring. Other tillage was with a sweep implement.
§ Average amount of residues at start of fallow was 6.6 Mg ha^{-1}. Values given represent percentage of original amount remaining.
¶ Determined to a 3-m depth.

Table 5-6. Progress in fallow systems with respect to water storage and wheat yields, Akron, CO (Greb, 1979).

Years	Tillage during fallow†	Fallow water storage		Wheat yield
		mm	%	Mg ha^{-1}
1916-1930	Maximum tillage; plow harrow (dust mulch)	102	19	1.07
1931-1945	Conventional tillage; shallow disk, rod-weeder	118	24	1.16
1946-1960	Improved conventional tillage; began stubble mulch in 1957	137	27	1.73
1961-1975	Stubble mulch; began minimum tillage with herbicides in 1969	157	33	2.16
1976-1990	Projected estimate; minimum tillage; began no-tillage in 1983	183	40	2.69

† Based on about 425 d of fallow (from mid-July to second mid-September).

Table 5-7. Straw mulch effects on soil water storage efficiency at Sidney, MT; Akron, CO; and North Platte, NE, from 1962 to 1965 (Greb et al., 1967).

Mulch rate	Fallow period precipitation	Water storage efficiency
Mg ha^{-1}	mm	%
0	355	16
1.7	355-549	19-26
3.4	355-648	22-30
6.7	355-648	28-33
10.1	648	34

Table 5-8. Straw mulch effects on soil water storage during fallow†, water storage efficiency, and grain sorghum yield at Bushland, TX, 1973 to 1976 (Unger, 1978).

Mulch rate	Water storage‡	Storage efficiency§	Grain yield
$Mg\ ha^{-1}$	mm	%	$Mg\ ha^{-1}$
0	72c*	22.6c	1.78c
1	99b	31.1b	2.41b
2	100b	31.4b	2.60b
4	116b	36.5b	2.98b
8	139a	43.7a	3.68a
12	147a	46.2a	3.99a

* Column values followed by the same letter are not significantly different at the 5% level (Duncan's multiple range test).
† Fallow duration of about 305 to 335 d.
‡ Water storage determined to a 1.8-m depth. Precipitation averaged 318 mm.

CO, as influenced by changes in management practices over time, is illustrated in Table 5-6. The approximate doubling of water storage during fallow and of yields is attributed to better weed control and to maintenance of surface residues, which improved infiltration and suppressed evaporation. Undoubtedly, improved varieties, fertility, and use of growing-season precipitation also were involved. However, results by Greb et al. (1967) and Unger (1978) clearly illustrate the benefits of surface residues for conserving water (Tables 5-7 and 5-8) and increasing crop yields (Table 5-8).

CURRENT RESIDUE MANAGEMENT SYSTEMS

Except for the stubble mulch tillage system, there was limited interest in crop production systems involving surface residues until the late 1960s or early 1970s when interest in conservation tillage became widespread. Conservation tillage, by most definitions, is a crop production system involving the management of surface residues. Certainly, studies conducted from the 1930s to the 1960s provided important background information for the currently used residue management systems, which hereafter are called *conservation tillage systems.*

Reasons for current interest in conservation tillage vary from region to region. One major reason is its effectiveness for controlling erosion. Closely allied are the water conservation benefits, not only in semiarid and subhumid regions but also in humid regions.

Soil and water conservation per se often have limited appeal to producers unless economic advantages result from their use. Economic factors contributing to interest in conservation tillage include: (i) high costs of fuel, labor, tractors, and other equipment; (ii) high equipment inventories and maintenance costs; (iii) ability to use erosive land for more intensive crop production (rather than for pastures or in long-term rotations); (iv) opportunities for more intensive cropping, rather than use of systems involving a long fallow, because of greater water conservation; and (v) in many instances, higher crop yields.

Other factors include the development and increased availability of improved herbicides, development of equipment suitable for use under surface residue conditions, opportunities for less soil compaction, and the realization that intensive tillage may not be essential for successful crop production and that it increases potential soil erodibility (Black and Siddoway, 1979).

Great Plains

Much of the early residue management research related to water conservation was conducted in the Great Plains. Therefore, it is not surprising that much current research regarding conservation tillage in the Great Plains is aimed at improving water conservation and subsequent use of that water.

Northern Great Plains

Besides evaporation suppression and weed control, water conservation in the Northern Great Plains generally involves snow management because snow may constitute a significant portion of the precipitation. However, water conservation during seasons other than winter is important also. The results for different segments (harvest to first winter, overwinter to spring, spring to fall, second overwinter to spring) of two common cropping systems (spring grain-fallow and winter wheat-fallow) will be discussed briefly. The last segment applies only to the spring grain-fallow system.

Water conservation during late summer depends on the crop grown and is greater after crops such as spring wheat and barley (*Hordeum vulgare* L.), which are harvested in late July or early August, than after sunflower (*Helianthus annuus* L.), which uses water well into September (Black and Bauer, 1985). Besides affording little time for storing water, sunflower residues have little effectiveness for water conservation purposes because of limited surface cover. The amount of residues available and precipitation amount and distribution affect the amount of water conserved. Generally, more than about 2.5 Mg ha^{-1} of small grain residues are needed to suppress evaporation during summer periods. Sufficient precipitation to wet the soil to depths greater than about 10 cm aids in suppressing evaporation (Black and Bauer, 1985).

Standing small grain stubble effectively traps snow. In Saskatchewan, soil water storage was 51 mm with standing stubble and 11 mm with bare fallow (Staple et al., 1960). In northeastern Montana, overwinter water storage was 76 mm with standing (no-tillage) spring wheat stubble (30-cm high) and 48 mm where the stubble was disked in the fall (Black and Power, 1965). In North Dakota, overwinter soil water storage ranged from 32 to 90 mm with no fall tillage and from 5 to 54 mm after fall moldboard plowing (Bauer and Kucera, 1978). Water storage with disking and V-blade undercutting was intermediate compared to the above treatments.

Besides standing stubble per se, stubble height affects water conservation (Black and Siddoway, 1977) (Table 5–9), with amount of water storage being closely related to the water equivalent of trapped snow. Similar results

Table 5-9. Snow depth, water equivalent of snow, and total soil water content as affected by stubble management (Black and Siddoway, 1977).

Date	Stubble height, cm			
	0†	15	28	38
		Snow depth, mm		
2 Apr. 1975	61	119	198	259
		Water equivalent of snow, mm		
2 Apr. 1975	13	25	43	56
		Total soil water content, mm		
12 May 1975	267	277	287	307

† This treatment involved conventional disk tillage.

were obtained by Aase and Siddoway (1980). Further improvements in snow trapping occurred when strips of short and tall stubble were alternated. This increased snow depth and density (resistance to wind passage) and increased overwinter water storage about 30% over that obtained with uniform medium-height stubble (Nicholaichuk et al., 1986; Willis and Frank, 1975).

Another improved technique of snow trapping is the use of vegetative barriers. Near Sidney, MT, tall wheatgrass [*Agropyron elongatum* (Host) Beauv.] barriers (two rows 0.9 m apart) established at 9.1- or 18.2-m intervals increased average snow depth about 38 and 20 cm, respectively, as compared with that on areas without barriers. The extra snow contributed about 12.7 and 7.6 mm more water to the soil, respectively, than the check (Greb and Black, 1971). Higher soil water contents and wheat yields with than without barriers were reported also by Aase and Siddoway (1976). The wheatgrass barriers slowed soil drying; thus improving germination, seedling emergence and establishment, and resistance to wind erosion.

For the spring to fall (summer of fallow) period, soil water storage usually is low (Black and Bauer, 1985). Whereas storage efficiencies during the overwinter period ranged from 60 to 84%, maximum storage during summer was 25% and water losses (11%) occurred in some cases. In general, additional water storage is low when the soil contains more than 75 mm of water at the start of the summer period (Black and Bauer, 1985). Conditions essential for additional water storage during the summer period are (i) an unfilled water storage reservoir, (ii) precipitation sufficient in amount and frequency to cause water to move deep enough into the soil to minimize evaporation, and (iii) sufficient crop residues to suppress evaporation. If the water content is high initially, immediate cropping to maximize water conservation and use would be more appropriate than continuation of fallow.

Unless precipitation is adequate to move water deep into the soil, most water will be lost by evaporation during summer. Surface residues help reduce evaporation. However, while they reduce the rate, they do not necessarily reduce total evaporation. For field conditions in Montana, at least 2.5 Mg ha^{-1} of wheat residues are needed to obtain significant increases in water conservation with no-tillage over that obtained with stubble mulch tillage (Tanaka, 1985).

As for the summer, water storage during the second overwinter period may be slight, depending on storage that occurred previously. A major concern during this period is wind erosion control, and conservation tillage has major potential for reducing wind erosion.

Central Great Plains

Fallowing is widely used for winter wheat in the Central Great Plains, with a 2-yr wheat-fallow system being used extensively. A 3-yr system of wheat-corn (or sorghum or millet)-fallow also is becoming popular and widely accepted.

A major goal of fallowing is to adequately recharge the soil profile with water so that the risk of failure for the next crop is greatly reduced. For three locations, wheat yields after fallow (one crop in 2 yr) were more than double those of continuous wheat (Greb et al., 1974).

Effective fallowing for water conservation involves weed control throughout the fallow period and retention of standing crop stubble to trap and retain snow and to minimize water loss by evaporation during the warm windy season. Because of earlier wheat (or other small grain) harvest in the Central than in the Northern Great Plains, there is greater opportunity for water storage from harvest to winter in the Central Great Plains. An important factor for water conservation during this period is weed control. Favorable weed control and water conservation are possible with stubble mulch tillage. Even greater water conservation has been achieved where herbicides were used alone or in combination with tillage to control weeds (Tables 5-3, 5-4, and 5-5). Besides favorable weed control, greater amounts of surface residues maintained with herbicide or herbicide-tillage treatments as compared with tillage alone also contributed to the greater water conservation.

The effects of surface residues on runoff and infiltration have been discussed previously. Residues reduce surface temperatures (Table 5-10) and, thereby, decrease evaporation because of lower vapor pressure of soil water. Residues also increase the thickness of the relatively nonturbulent air layer at the soil-air interface, which decreases the rate of vapor transport away from the soil. Smika (1983) demonstrated the latter under field conditions by placing equal amounts of wheat straw (4.6 Mg ha^{-1}) on soil in different

Table 5-10. Average daily soil-surface temperature as affected by bare soil and wheat straw position during 5 weeks (August-September) at Akron, CO (Smika, 1983).

Straw position†	Soil-surface temperature‡
	°C
Bare soil	47.8c*
Flat straw	41.7b
3/4 flat, 1/4 standing	39.6b
1/2 flat, 1/2 standing	32.2a

* Column values followed by the same letter are not significantly different at the 1% level (Duncan's multiple range test).

† All straw amounts were 4.6 Mg ha^{-1}.

‡ Average of measurements at 1000, 1200, and 1500 h with a radiation thermometer.

positions. Windspeeds needed to initiate water loss increased as the amount of standing stubble increased. Also, water loss rates decreased with increasing amounts of standing stubble at a given windspeed.

After the surface dries, water flow to the surface and air permeability of the surface soil become important factors in the evaporation process. This was demonstrated by Smika (1976), who compared evaporative losses from conventional-, minimum-, and no-tillage areas during a 34-d period after 165 mm of rainfall. On the day after rainfall, water contents were similar for all treatments (Fig. 5–1a). After 34 rainless days (Fig. 5–1b), soil water content dropped to 0.1 $m^3 m^{-3}$ to depths of 12, 9, and 5 cm with conventional-, minimum-, and no-tillage treatments, respectively. Depths of major drying for conventional- and minimum-tillage treatments were similar to the depth (10 cm) at which blade tillage was performed 8 d before rainfall. Surface residues with minimum- and no-tillage treatments reduced evaporation as compared with conventional tillage. Amounts of surface residue present during the 34-d period were 1.2, 2.2, and 2.7 Mg ha^{-1} with conventional-, minimum-, and no-tillage treatments, respectively.

As in the Northern Great Plains, snow trapping is important for water conservation in the Central Great Plains. Average overwinter water storage with standing stubble was 51 mm, which represented 99% of the overwinter precipitation received. Such high value indicates that trapping of some snow from adjacent areas had occurred. Where stubble was incorporated by oneway

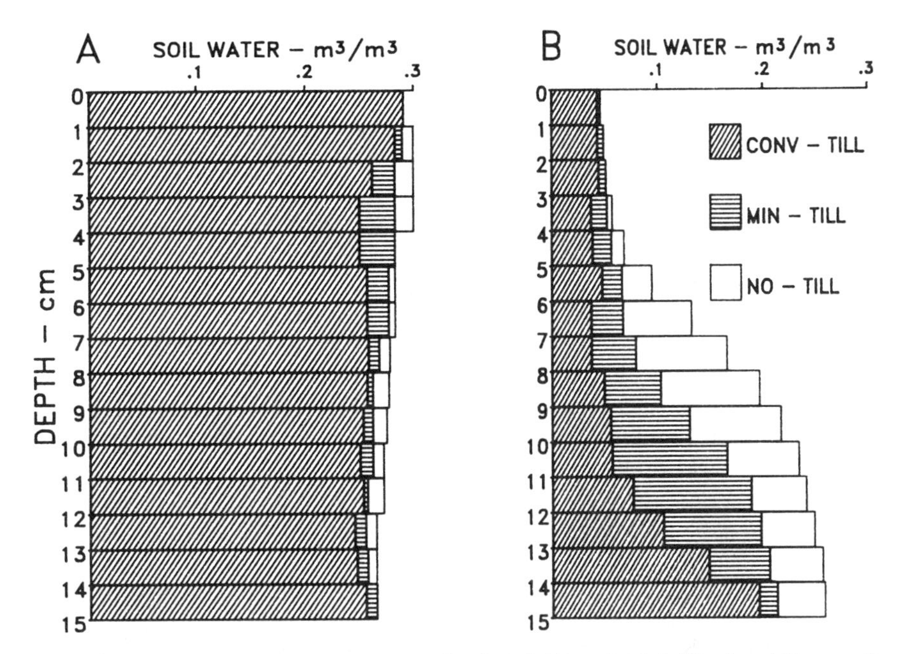

Fig. 5–1. Soil water contents to a 15-cm depth 1 d (A) and 34 d (B) after 165 mm of rainfall as influenced by tillage treatments (CONV-TILL, conventional tillage; MIN-TILL, minimum tillage; NO-TILL, no-tillage) (Smika, 1976).

disk tillage, a soil water loss of 3 mm occurred during the same period (Smika and Whitfield, 1966).

Barriers of tall wheatgrass, sorghum, or sudangrass [*S. sudanense* (Piper) Stapf.] effectively trap snow in the Central Great Plains (Greb and Black, 1971). For maximum effectiveness, vegetative barriers should grow about 0.4 to 1.2 m tall, have flexible stalks, and produce 25 to 35% density. Additionally, the barriers should be perpendicular to prevailing winds during snowstorms and be spaced 9 to 18 m apart to coincide with equipment used to farm the land (Smika, 1985). At Akron, CO, vegetative barriers increased water storage by 38 mm and winter wheat grain yields by 0.27 Mg ha^{-1}. They also helped to control wind erosion (Smika, 1985).

Southern Great Plains

Water conservation undoubtedly is more difficult in the Southern than in the Central and Northern Great Plains. The greater difficulty results from such factors as higher potential evaporation, mainly late spring to early fall precipitation (during warm season when evaporation is high), highly variable precipitation, a long growing season for weeds, and low levels of crop residues associated with generally low crop yields (under dryland conditions). Low water infiltration rates contribute to low water conservation on some soils.

Water content and crop yield differences with sweep- and no-tillage treatments on dryland in early studies and in a more recent study (O.R. Jones, 1987, personal communication) generally have been slight because of low residue production. However, greater differences between no-tillage and sweep- or disk-tillage treatments have been obtained when residues from irrigated crops were involved. During a fallow period from winter wheat harvest in July 1968 until sorghum planting in May 1969, weeds and volunteer wheat were controlled with combinations of disk, sweep, and herbicide treatments. The wheat had produced 11 Mg of residue ha^{-1}. Water storage with treatments involving herbicides (atrazine and 2,4-D), either alone or after one sweep-tillage operation, was about double the average for treatments involving only tillage (disk, sweep, or disk plus sweep) (Unger et al., 1971). Sorghum yields were not determined.

In a WSF study involving irrigated winter wheat and dryland grain sorghum, Unger and Wiese (1979) used no-, sweep-, and disk-tillage methods to manage wheat residues and control weeds and volunteer wheat during the fallow period between wheat and sorghum (Table 5-11). Water storage, water use, grain yields, and water-use efficiency were highest with no-tillage. For an irrigated wheat-dryland sorghum-dryland sunflower rotation study, water storage and sorghum yields again were highest with no-tillage (Table 5-12) (Unger, 1984). Total water use by sorghum in both studies was greatest with no-tillage, but that was more than offset by higher water contents at sorghum planting. This suggests that water contents were higher with no-tillage throughout the growing season, thereby allowing no-tillage sorghum to respond more favorably to growing-season precipitation. Analyses by Unger et al. (1986) showed that yields of grain sorghum increased in response to growing-season precipitation as amounts of surface residue increased. The

Table 5-11. Tillage effects on water storage, sorghum grain yields, and water-use efficiency (WUE) in an irrigated winter wheat-fallow-dryland grain sorghum cropping system, Bushland, TX, 1973 to 1977 (Unger and Wiese, 1979).

Tillage method	Plant-available water at sorghum planting	Water storage efficiency†	Grain yield	Total water use	WUE‡
	mm	%	Mg ha^{-1}	mm	kg m^{-3}
No-tillage	217a*	35.2a	3.14a	350a	0.89a
Sweep	170b	22.7b	2.50b	324b	0.77b
Disk	152c	15.2c	1.93c	320b	0.66c

* Column values followed by the same letters are not significantly different at the 5% level (Duncan's multiple range test).
† Precipitation averaged 347 mm during fallow.
‡ Water-use efficiency based on grain yields, growing season precipitation, and soil water changes.

response was greatest during the vegetative growth stage (Table 5-13, Eq. [6], [7], and [8]).

Annual cropping with full irrigation usually results in highest total crop production, but water for irrigation is limited in the Southern Great Plains. Consequently, alternatives to full irrigation have been sought. One approach is the use of conservation tillage systems to more effectively conserve and use water from precipitation for crop production. When Musick et al. (1977) used no-tillage and clean (disk)-tillage methods during fallow after wheat in a WSF rotation, precipitation storage efficiencies were 35 and 21%, respectively, on level-bordered plots and 47 and 28%, respectively, on graded-furrow plots. Sorghum grain yields on level plots averaged 5.10 and 4.08 Mg ha^{-1} with the respective treatments with 150 mm of irrigation. With 300 mm of irrigation, respective yields were 6.46 and 5.97 Mg ha^{-1}. On graded-furrow plots, 169 mm of irrigation water was retained with no-tillage, and sorghum yielded 5.42 Mg ha^{-1}. With clean tillage, 93 mm of irrigation water

Table 5-12. Effect of tillage method on average soil water storage during fallow after irrigated winter wheat and on subsequent dryland grain sorghum yields at Bushland, TX, 1978 to 1983 (Unger, 1984).

Tillage treatment	Precipitation		Water storage		Grain yield	Total water use	WUE‡
	Fallow†	Growing season					
	mm		mm	%§	Mg ha^{-1}	mm	kg m^{-3}
Moldboard	316	301	89b*	29b	2.56bc	360bc	0.71
Disk	316	301	109b	34ab	2.37cd	363bc	0.65
Rotary	316	301	85b	27b	2.19d	357c	0.61
Sweep	316	301	114ab	36ab	2.77b	386ab	0.72
No-tillage	316	301	141a	45a	3.34a	401a	0.83

* Column values followed by the same letter or letters are not significantly different at the 5% level (Duncan's multiple range test).
† Fallow duration about 305 to 335 d.
‡ Water-use efficiency (WUE) based on grain yield, growing season precipitation, and soil water changes.
§ Based on fallow period precipitation stored as soil water.

Table 5–13. Sorghum yield response to precipitation with different amounts of wheat residue on the soil surface at sorghum planting time (Unger et al., 1986).

Variables		Residue				Coefficient of determination, r^2	
Dependent	Independent	level	Observation	Equation	Equation	Value	Significance level
		Mg ha^{-1}	no.				
Grain yield, Mg ha^{-1}	Growing season precipitation, mm	0–8.0	399	1	y = 0.941 + 0.0066x	0.116	0.0001
		0–0.4	117	2	y = 1.041 + 0.0045x	0.145	0.0001
		0.5–3.2	156	3	y = 0.961 + 0.0069x	0.145	0.0001
		>3.2	126	4	y = 0.586 + 0.0092x	0.078	0.0015
	Vegetative period precipitation, mm	0–8.0	399	5	y = 1.032 + 0.0203x	0.307	0.0001
		0–0.4	117	6	y = 1.050 + 0.0144x	0.357	0.0001
		0.5–3.2	156	7	y = 1.103 + 0.0199x	0.273	0.0001
		>3.2	126	8	y = 0.866 + 0.0271x	0.370	0.0001
	Heading period precipitation, mm	0–8.0	399	9	y = 1.166 + 0.0151x	0.182	0.0001
		0–0.4	117	10	y = 0.871 + 0.0134x	0.297	0.0001
		0.5–3.2	156	11	y = 1.207 + 0.0151x	0.235	0.0001
		>3.2	126	12	y = 1.178 + 0.0195x	0.148	0.0001
	Grain-filling precipitation, mm	0–8.0	399	13	y = 2.800 − 0.0049x	0.047	0.0001
		0–0.4	117	14	y = 2.295 − 0.0046x	0.068	0.0045
		0.5–3.2	156	15	y = 2.838 − 0.0040x	0.033	0.0239
		>3.2	126	16	y = 3.390 − 0.0079x	0.101	0.0003

was retained and yields averaged 4.26 Mg ha^{-1}. Higher yields with no-tillage resulted from greater water storage during fallow and greater irrigation water infiltration. The latter occurred even though no-tillage plots contained more water than clean-tillage plots at planting time.

Use of conservation tillage for annually irrigated crops often is difficult because of planting problems in heavy residues and because of poor weed and volunteer plant control. These problems may be especially severe when wheat, corn, and grain sorghum are grown continuously. In such cases, limited rather than no-tillage methods generally have given better results.

Allen et al. (1976) used no-, limited-, and clean-tillage methods for furrow-irrigated winter wheat at Bushland, with no- and limited-tillage treatments being alternated annually. Herbicides satisfactorily controlled weeds and volunteer wheat on no-tillage plots in 2 yr, but a second contact herbicide application was needed the 3rd yr when rainfall was above average. On limited- and clean-tillage plots, herbicides and/or tillage satisfactorily controlled weeds and volunteer wheat. Seeding and plant establishment were satisfactory with a conventional drill each year except on no-tillage plots in 1 yr when large amounts of residue (about 10 Mg ha^{-1}) caused limited disk-opener penetration. Irrigation water advanced satisfactorily in no-tillage furrows. Average infiltration was higher with clean- than with limited- or no-tillage treatments, both with adequate- and limited-irrigation. Grain yields were significantly higher with no- than with clean-tillage with limited irrigation (2.91 vs. 2.61 Mg ha^{-1}). Yield differences were not significant with adequate irrigation (3.31 vs. 3.02 Mg ha^{-1}). For both irrigation levels, irrigation water use efficiency was significantly higher with no- than with clean-tillage. Although favorable results were obtained with no-tillage, Allen et al. (1976) considered alternating between limited and no-tillage more practical and dependable for continuous irrigated wheat production because of less severe problems than with no-tillage.

Excessive volunteer sorghum has decreased yields of continuous irrigated grain sorghum under no-tillage conditions (Allen et al., 1975a). "Safened" seed that allows use of herbicides to control the volunteer sorghum has shown some promise, but limited tillage generally is more satisfactory. Allen et al. (1980) showed that a mulch-subsoil (limited tillage) treatment involving anhydrous ammonia application in furrows by subsoiling deep in fall followed by sweep-rodweeding and planting in spring significantly increased irrigation water infiltration (386 vs. 347 mm) and grain yields (5.92 vs. 5.16 Mg ha^{-1}) over that with clean tillage (disk, chisel, bed, sweep-rodweed, plant) with limited irrigation. Infiltration (483 vs. 437 mm) and yields (6.86 vs. 6.35 Mg ha^{-1}) were similar with adequate irrigation. Water-use efficiency differences due to tillage were similar with both irrigation levels.

For irrigated grain sorghum double-cropped after irrigated wheat, no-tillage sorghum emerged sooner, grew taller, and matured up to 5 d earlier than clean-tillage sorghum. Grain yields for the 5-yr study averaged 5.69 and 5.07 Mg ha^{-1} with no- and clean-tillage treatments, respectively. Water-use efficiency was higher with no-tillage because of the higher yields and no differences in total water use (Allen et al., 1975b).

In contrast to the semiarid portion of the Southern Great Plains where precipitation averages only about 470 mm annually (Bushland area), precipitation in the subhumid and humid regions of the Southern Great Plains states (Texas and Oklahoma) ranges to slightly more than 800 mm annually. Although water conservation is important also in those regions, conservation tillage often had no effect on soil water storage in subhumid parts of Texas (Clark, 1983; Gerik and Morrison, 1984, 1985; Matocha and Bennett, 1984). Yields (wheat, grain sorghum, and corn) usually were not affected but were lower with no-tillage in some cases because of greater weed problems and planting problems under high-residue conditions (Gerard and Bordovsky, 1984; Unger et al., 1987).

As in Texas, water conservation in Oklahoma is more important in semiarid regions than in more humid regions. However, in contrast to results in Texas, substantial increases in soil water storage and reductions in evaporation have occurred in the more humid regions of Oklahoma where crop residues were maintained on the soil surface. In addition, wheat yields were higher with reduced than with conventional tillage (3.76 vs. 3.11 Mg ha^{-1}) (Stiegler et al., 1984).

Midwest and Northeast

Conservation tillage studies in the Midwest and Northeast generally concern the effects of surface residues on controlling erosion, mainly by water. Any water conservation benefits result primarily from increased water infiltration and decreased runoff. These factors may or may not increase soil water content, depending on antecedent soil water content. However, they have potential for more readily filling the soil with water and, therefore, may be important during periods of limited or infrequent precipitation.

Runoff generally was less with conservation than with clean tillage (Andraski et al., 1985a, 1986b; Griffith et al., 1977; Laflen et al., 1978), with surface residue amount and surface coverage provided by residues being important factors. Other factors influencing runoff were antecedent rainfall; rainfall amount, intensity, and energy; plant cover; and soil properties. In addition, the type of conservation-tillage system used affected runoff under some conditions. For example, Andraski et al. (1985b) showed that chiseling was more effective than till-planting or no-tillage for reducing runoff when water was applied soon after planting. At later times, runoff differences for the conservation-tillage methods were not significant.

During a major rainstorm in Ohio, runoff from a no-tillage watershed was 64 mm as compared with 58 mm from a clean-tilled watershed with rows on the contour and 112 mm from a clean-tilled watershed with sloping rows (Harrold and Edwards, 1972). Rainfall was 129 mm on the no-tillage watershed and 140 mm on the other watersheds. Although rainfall was 11 mm less on the no-tillage watershed, the slope was much greater (20.7% vs. 5.8 and 6.6% on the contour- and sloping-row watersheds, respectively). All watersheds were planted to corn. Soil loss was 0.07 Mg ha^{-1} with no-tillage and 50.7 and 7.2 Mg ha^{-1} with sloping- and contour-row clean-tillage watersheds, respectively.

In Pennsylvania, McClellan (1970) obtained greater water infiltration and conservation with no-tillage than with conventional tillage for corn production after sod. For doublecropping in Maryland, Parochetti (1970) reported that no-tillage conserved water that would normally be lost during tillage operations for the next crop.

In the above studies, higher soil water contents were implied but the amounts generally were not reported. Johnson et al. (1984), however, showed that average water contents at corn planting time were higher with three conservation-tillage systems (chisel plowing, till-planting, and no-tillage) than with conventional moldboard plowing. For most of the growing season, the water content was highest with no-tillage. However, during a period of heavy rainfall (93-mm total), chiseling resulted in greatest recharge of the soil profile. The study was conducted in Wisconsin, on a Griswold silt loam (fine-loamy, mixed, mesic Typic Argiudoll) having a 6% slope.

Although runoff generally decreases with increases in amounts of surface residues, surface residues alone may not reduce runoff if soil conditions are not favorable for rapid water infiltration. For example, Lindstrom and Onstad (1984) showed that with continuous corn, runoff from simulated rainfall was greater for no-tillage that provided 64% surface coverage by residues than for reduced tillage (fall chisel plow, spring disk) with 33% surface cover or conventional tillage (fall moldboard plow, spring disk) with only a trace of surface residues. The measurements were made during the 3rd crop yr. Increased runoff with no-tillage was attributed to soil surface with a high bulk density, high-penetrometer resistance, low-saturated hydraulic conductivity, and low volume of macropores. Cultivation of the crop on no-tillage plots during the growing season to establish ridges for planting of the next crop probably contributed to the poor soil physical conditions and, subsequently, the high runoff. The study was conducted on a Barnes loam (fine-loamy, mixed Udic Haploborall) near Morris, MN.

On Nicollet clay loam (fine-loamy, mixed, mesic Aquic Hapludoll) near Waseca, MN, Lindstrom et al. (1981) obtained higher infiltration rates of simulated rainfall with conventional (fall moldboard, spring field cultivate) and conservation (fall chisel plow, spring field cultivate) tillage than with no-tillage for nonwheel-tracked areas. For wheel-tracked areas, differences in infiltration rates among treatments were slight. The tillage treatments had been continuously applied to the same area for 10 yr before making the determinations. Lower infiltration on nonwheel-tracked areas with no-tillage was attributed to lower random roughness and higher bulk density than with other treatments. On tracked areas, differences among treatments for bulk density and random roughness were smaller than on nontracked areas but remained higher with no-tillage. The generally less-favorable soil conditions with no-tillage probably developed before complete surface coverage was provided by residue or even before the no-tillage system was established (Lindstrom et al., 1981). The results indicate that no-tillage may not be the best treatment in all cases, even though it maintained the greatest amount of residues on the surface. Also, no-tillage alone may not alleviate an undesirable soil condition.

South and Southeast

The thermic-humid climate of the south and southeast USA permits intensive cropping. However, despite favorable annual precipitation totals, droughts of varying duration (but generally short-term) may severely limit crop yields where soils have low water-storage capacity and high-intensity rainfall results in excessive runoff and erosion (Reicosky et al., 1977). Contributing to low infiltration and high runoff in some cases are steep slopes and soils with low water retention potential and with genetic or traffic-induced hardpans at relatively shallow depths (Langdale et al., 1981; Sojka et al., 1984). Consequently, major reasons for using conservation tillage are to control erosion, increase infiltration, reduce evaporation, minimize soil strength (by maintaining more favorable soil water contents), and increase the potential for multiple cropping by compressing the timetable for field operations (Sojka et al., 1984).

The effects of conservation tillage on water conservation and use in the South and Southeast were included in reviews by Phillips et al. (1980), Reicosky et al. (1977), and Sojka et al. (1984). In general, water conservation and crop yields were as great or greater with conservation tillage than those with clean tillage when residues of previous crops or cover crops were maintained on the surface, provided that adequate nutrients were available and that sod was killed long enough before planting to avoid competition for water. When planted in stunted or live sod, yields usually were lower with conservation tillage than with plowing unless the crop was irrigated and adequately fertilized. Satisfactory weed control was necessary in all cases.

In central Kentucky, average water contents to a 15-cm depth in Maury silt loam (fine, mixed, mesic Typic Paleudalf) were 0.295 and 0.244 $m^3\ m^{-3}$ with no- and conventional tillage, respectively, during the growing season for corn. Transpiration by corn and soil water evaporation for the tillage systems are given in Table 5–14. Total transpiration was 65 mm greater with no-tillage than with conventional tillage. Evaporation was 150 mm greater

Table 5–14. Estimated soil water evaporation and transpiration by no-tillage and conventional tillage corn crops on Maury silt loam soil, 1970 to 1973 average (Phillips et al., 1980).

	No-tillage		Conventional tillage		
Month	Transpiration	Evaporation	Transpiration	Evaporation	Rainfall
	mm				
May	0	21	0	63	179
June	76	10	64	68	97
July	124	3	95	21	101
August	92	2	72	14	41
September	15	5	11	25	91
Total	307	41	242	191	509

with conventional tillage. Corn yields obtained with the two tillage systems on several soils are given in Table 5-15. The weighted average was 9.24 Mg ha^{-1} with no-tillage and 8.78 Mg ha^{-1} with conventional tillage (Phillips et al., 1980).

Because year-round soil protection is essential for controlling erosion and year-round production is desirable for an efficient farming enterprise, considerable research has been conducted involving grain crop planting in live or stunted sods and sod reestablishment after grain crop harvest. With adequate rainfall, yields of corn planted in live sod were not affected by tillage method (conventional-, reduced-, or no-tillage). Irrigation was required for good yields when rainfall was limited (Adams et al., 1970; Beale and Langdale, 1964; Carreker et al., 1972). Forage yields of the sod crop after corn depended on the degree of competition between the corn and sod crops, but generally were substantially higher where corn had been no-tillage planted in the sod (Adams et al., 1970; Beale and Langdale, 1964), apparently because of higher soil water contents. Higher surface soil water contents (0–5 cm depth) after no-tillage corn harvested for silage also aided reestablishment and early growth of orchardgrass (*Dactylis glomerata* L.) and red clover (*Trifolium pratense* L.). The higher water contents resulted from the surface residues that remained from the grass-clover crop that preceded the no-tillage corn (Moschler et al., 1969).

Conservation tillage technology to use and manage crop residues in the thermic-humid South and Southeast has been slowly emerging during the past 40 yr. Some relatively recent major developments include the introduction of in-row subsoiling-planting equipment, doublecropping systems, and practices to restore productivity of severely eroded soils.

Because of the dense hardpan of Ultisols (B2t or E horizons) that restricts water infiltration and root growth, subsoiling is a common practice on some soils of the South and Southeast. Disrupting the hardpan increases the volume of soil that can be used to store water and increases root proliferation to use that water, thus minimizing the potential adverse effects of short-term droughts on crop yields. With conventional tillage, however, benefits from subsoiling are relatively short-lived because subsequent tillage and precipitation on the unprotected surfaces move soil into the subsoiler slots, thus decreasing their effectiveness for improving infiltration and root prolifera-

Table 5-15. Average grain yields of corn grown under no-tillage and conventional tillage conditions on several well-drained soils in Kentucky (Phillips et al., 1980).

Soil type and classification	Years tested	Grain yield: No-tillage	Grain yield: Conventional tillage
	no.	Mg ha^{-1}	
Maury (fine, mixed, mesic Typic Paleudalf)	8	9.14	8.93
Crider (fine-silty, mixed, mesic Typic Paleudalf)	5	9.89	8.32
Tilsit (fine-silty, mixed, mesic Typic Fragiudult)	5	7.71	7.71
Allegheny (fine-loamy, mixed, mesic Typic Hapludult)	3	10.98	10.91
Weighted avg.		9.24	8.78

Table 5-16. Effect of tillage, crop cover, and initial soil water content on runoff of simulated rainfall† and soil loss (Langdale et al., 1973).

Tillage, cover, initial water content	Soil slope, %			
	10.7		21.4	
	Runoff	Soil loss	Runoff	Soil loss
	%	Mg ha^{-1}	%	Mg ha^{-1}
Subsoiled, rye mulch				
5.8%	4	0.05	13	0.3
14.2%	37	0.4	67	0.7
Subsoiled, soybean canopy (68%) over rye mulch				
8.7%	19	0.1	44	0.1
14.1%	62	0.2	81	0.2
Tilled, bare fallow				
12.0%	72	36.0	85	39.0
14.7%	88	39.0	92	50.0

† Applied at 63.5 mm h^{-1} for 2 h.

tion. In contrast, in-row subsoiling-planting equipment makes it possible to retain essentially all existing plant residues on the soil surface to protect the slots and, thereby, obtain longer-lasting benefits (Sojka et al., 1984). On Cecil sandy loam (clayey, kaolinitic, thermic Typic Hapludult) in Georgia, runoff and soil losses were lower with in-row subsoiling-planting than with a tilled-bare fallow treatment (Table 5-16). Besides reducing runoff and soil losses, in-row subsoiling-planting of soybean [*Glycine max* (L.) Merr.] in rye (*Secale cereale* L.) cover reduced losses of some nutrients (Langdale et al., 1973).

The thermic-humid climate in the South and Southeast makes doublecropping possible. When grain crops are grown, economic returns generally are greater than when grain and forage crops are rotated. Langdale et al. (1984) evaluated conventional (disk harrow in fall and spring), minimum (disk harrow in fall), and no-tillage treatments for wheat and grain sorghum in Georgia. Wheat grain yields averaged 3.08, 3.08, and 3.13 Mg ha^{-1} for the respective treatments, but the differences were not significant. Sorghum grain yields were significantly higher with minimum and no-tillage (3.72 and 3.73 Mg ha^{-1}, respectively) than with conventional tillage (2.88 Mg ha^{-1}). Total stover yields (wheat plus sorghum) averaged over 11.0 Mg ha^{-1} with minimum and no-tillage treatments and were manageable by both conservation tillage methods. However, minimum tillage may be more practical than no-tillage for long-term doublecropping because no-tillage could lead to such problems as soil compaction in wheel tracks and increased competition by weeds. In addition, soil fertility was best managed with minimum tillage. Because of the large amounts of residue produced, doublecropping of grain crops is effective for controlling erosion (Langdale et al., 1979, 1984; Mills et al., 1986).

Many areas of the South and Southeast have experienced major soil losses by erosion for many years. To restore soil productivity, use of perennial legumes has been widely investigated, generally in long-term cropping systems having low economic value. Recent research (Langdale et al., 1987) has shown that an annual cool-season legume (crimson clover—*Trifolium incarnatum* L.) provided adequate N and surface mulch so that yields of rain-fed grain sorghum were higher with no- than with conventional tillage where the erosion level was slight (3.95 vs. 3.70 Mg ha^{-1}) or moderate (5.71 vs. 3.90 Mg ha^{-1}). Where erosion was severe, yields were lower with no-tillage (3.95 vs. 5.02 Mg ha^{-1}). No-tillage sorghum received no N fertilizer, whereas 90 kg of N ha^{-1} were applied for conventional tillage sorghum. Under irrigated conditions, N applications were 146 and 56 kg ha^{-1} for conventional and no-tillage sorghum, respectively. Grain yields under irrigated conditions averaged 5.82, 5.71, and 6.52 Mg ha^{-1} with conventional tillage where erosion was slight, moderate, and severe, respectively. The respective yields were 5.51, 6.03, and 6.18 Mg ha^{-1} with no-tillage. Higher yields on more severely eroded soils were attributed to increased clover growth and subsequent buildup of soil organic matter, especially by the 3rd yr, because the clay content was highest where erosion was severe. The surface soil (0–7.5 cm depth) contained 6.0, 10.7, and 20.8% clay where erosion was slight, moderate, or severe, respectively. Eroded Southern Piedmont soils with slopes of about 11% or more often are not fully recharged with water when evapotranspiration is greater than rainfall (Bruce et al., 1987). Hence, surface mulches are important for water conservation on these soils (Langdale et al., 1987).

Pacific Northwest

The Pacific Northwest wheat region has humid winters and dry summers. The climate ranges from semiarid in the central areas to subhumid near the mountains that border the region. Water limits crop production in most dry-farmed areas. Yields depend on stored water because only 20 to 25% of the annual precipitation falls during the growing season. Precipitation occurs mostly as low-intensity rain, usually not exceeding 2 to 3 mm h^{-1}. Snowfall amounts to 15 to 20% of annual precipitation in northern areas and at higher elevations but decreases in the lower, drier areas. Soils often freeze to a depth of 40 cm or more, especially in the absence of snow cover. However, warm frontal systems or rainy weather commonly cause complete thaws several times during the winter.

The topography varies from nearly level valleys to steeply sloping uplands. Most steeper slopes are in the subhumid zone where some croplands have slopes of 50% or more. Soils are formed mainly from loess and are quite erodible. Most soils in the high-precipitation zone (400–600 mm) have a silt loam surface and silty clay loam or clay loam subsoils. Soils grade to sandy loams in the dry areas where annual precipitation ranges from 200 to 300 mm.

The climatic patterns, steep topography, and predominant winter wheat cropping creates a winter runoff and erosion problem in much of the region (Papendick and Miller, 1977). When soils are frozen, heavy runoff can occur, even with low intensity rains or melting snow. Early studies on frozen soil have shown that total runoff is closely correlated with snow cover and amount of rainfall because the infiltration rate is low (Horner et al., 1944).

For example, runoff from untilled stubble may exceed that from bare soil because the stubble holds more drifting snow. Runoff can also be large from the melt water of large snow drifts that accumulate on the steep north slopes.

Though runoff is significant, evaporation accounts for the greatest loss of water (Ramig et al., 1983). Estimates are that close to 75% of the annual precipitation can be lost by evaporation from a bare, uncropped soil (Bristow et al., 1986). However, with a typical wheat-fallow system, storage efficiencies range from 50 to 75% the first winter after crop harvest and 10 to 50% the second winter (Leggett et al., 1974). Efficiency is lower for the winter following wheat planting because the profile contains more water and there are less surface residues. This can result in more runoff and evaporation. On the average, fallow storage efficiency is about 50% for the period from September after harvest to March of the crop year (Ramig et al., 1983).

Several studies have shown that surface residues significantly increase overwinter water storage. At Lind, WA, 5 and 11 Mg ha^{-1} of wheat residues on the surface from September through March resulted in 30 and 50 mm more water storage, respectively, than where no residues were present (Papendick and Miller, 1977). Storage efficiencies were 48% for bare soil and 66 and 81% for the 5 and 11 Mg ha^{-1} mulch rates, respectively. However, oversummer evaporative losses were greater under the mulch than with bare soil and negated about half the gains that had occurred overwinter.

Ramig and Ekin (1978) showed that overwinter storage efficiency at Pendleton, OR was 87% for standing wheat stubble compared with 64% for fall plowing. This resulted in 66 mm of additional water, which increased subsequent green pea (*Pisum* spp.) production by 25% or nearly 0.70 Mg ha^{-1}. In north central Oregon, stubble significantly increased overwinter water storage compared with burned stubble, but there was little or no difference between standing or flailed stubble (Ramig and Ekin, 1984). Similarly, at Teton, ID, an additional 120 mm of water was stored with standing stubble as compared with stubble burned in the fall. Much of the additional water resulted from trapping drifting snow. Winter wheat yields were increased by about 1.1 kg ha^{-1} for each millimeter of trapped snow water (Massee and McKay, 1979).

Theoretical analyses and field experiences show that surface residues are of greatest value for water conservation during the winter rainy season and early spring when the soil surface is wet much of the time. The residue cover slows first-stage drying and, hence, allows more time for water to move deeper into the soil where it is less susceptible to loss by evaporation. An analysis of evaporation from untilled soil using a simultaneous heat-water flow simulation model showed that surface residues markedly suppressed the water loss rate during winter and early spring months but increased the loss rate during summer (Bristow et al., 1986). This occurred because the surface residues, by slowing the evaporation rate, extended the duration of first-stage drying (soil surface remained moist longer) and slowed advancement of second-stage drying compared with bare soil. Hence, for maximum water conservation, it is important to cultivate as soon as possible after major rains cease to break capillary continuity in the surface layers. This hastens the onset of second-stage drying in the tilled layer and sharply reduces the evaporation rate.

For most soils, residues have little or no beneficial effect for water conservation during the dry season compared with a dry-tillage mulch. At Pendleton, OR, water storage during summer was lower with chemical fallow (no-tillage fallow) than with tillage. Wheat yields also were lower because of drier fall seeding conditions (Oveson and Appleby, 1971; Swan et al., 1974). Similarly, early fall stand establishment was considerably more difficult without some fallow tillage because of greater oversummer loss of water from the seed zone in untilled soil, even though it was covered with some residues (Lindstrom et al., 1974).

Stubble mulching is the main surface residue management method used in wheat-fallow areas. It is accomplished with subsurface sweeps, chisel plows, or oneway disks for initial tillage, usually performed in early spring. A rodweeder or springtooth cultivator is used for secondary tillage to kill weeds and to form a tillage mulch while at the same time keeping the residues on the surface to control erosion. Farmers in more humid areas have been reluctant to practice stubble mulching because it depressed crop yields compared with clean tillage and tillage and planting equipment suitable for operation in heavy residues is lacking.

In recent years, there has been increased interest in no-tillage farming, particularly in the more humid areas and transitional areas between the subhumid and semiarid zones. Maximum retention of surface residues with no-tillage has permitted some farmers in the intermediate rainfall zones to increase cropping intensity (winter wheat followed by spring wheat or spring barley in lieu of winter wheat-fallow) because of improved water conservation. Results with no-tillage in the subhumid zone indicate that surface residues increase overwinter water storage during seasons of below normal precipitation but not in years when the soil profile is filled by early spring (Cochran et al., 1982). The additional water increased wheat yields and water-use efficiency unless there was a negating effect from increased grassy weeds associated with the surface residues.

A type of residue management termed *slot mulch* has been under development as a supporting practice to reduce runoff, especially from frozen soils (Saxton et al., 1981). Straw and chaff from the combine windrow are packed into a continuous slot 7 to 13 cm wide and 20 cm or more deep on the contour. The practical distance between slots would be the width of the combine, e.g., 6 to 7 m. The straw is left exposed at the surface, which maintains macroporosity for intercepting runoff and insulates the bottom of the slot so it remains largely unfrozen during cold weather. The slot mulch reduces runoff to almost zero for most precipitation events in the Pacific Northwest (Redinger et al., 1984). Slot mulching does not interfere with no-tillage double disk seeding but does with hoe-type openers. Commercial slot mulch machines are not yet available, but experimental types are being tested in eastern Washington.

Other new types of tillage-residue management methods under study to increase overwinter water storage include the paraplow and basin pitters. The paraplow is a slant-shank plow that is operated 30 to 40 cm deep to create fractures and large macropores in the soil, which help to maintain high infiltration rates, even when the soil is frozen. Preliminary research in a 350-mm precipitation area shows that 30 to 60 mm more water is stored overwinter with fall paraplowing as compared with no-tillage (L.F. Elliott et al., 1987, unpublished data). However, the soil water gains with either tillage or no-tillage were almost doubled when residues from the previous wheat crop were left intact compared with burning the residues.

Basin pitters form small pits or miniature basins that pond surface water during runoff events. The operation leaves the surface covered and enough stubble standing to trap blowing snow. Preliminary tests show that the pitter can be used on tilled and untilled stubble and shows promise as an economical practice for improving water conservation in the dryland areas.

RESEARCH NEEDS

Crop residue management or conservation tillage systems are widely recognized for their soil conservation benefits. They also effectively conserve water in many situations. However, much of the water that potentially could be conserved and used for crop production still is lost through runoff, evaporation, or transpiration by noncrop plants. The following are some examples of research that is needed to develop more effective and more widely adaptable conservation tillage systems.

1. More effective herbicides. Highly effective and specific herbicides are needed to control weeds and volunteer crop plants. The herbicides should provide control under a broad range of weed growth stages and environmental conditions but should not affect subsequent crops or harm the environment.
2. Improved equipment. Vast improvements in equipment capable of operation under high-residue conditions have been made in recent years, but equipment problems still exist. The improved equipment should provide trouble-free operation, provide effective separation between seeds and residues, cause minimum destruction of surface residues, and place seeds at a uniform depth.
3. Improved residue maintenance. Crop residues often are limited, especially in semiarid regions, and are subject to decay. If residues were more stable, a buildup of surface residues could occur. Research is needed to develop improved residue maintenance practices and to select for or develop cultivars having more stable residues. Further information on residue allelopathic effects is required also.
4. Improved cropping sequences. Research is needed to develop cropping sequences that result in more effective timing among periods of soil water storage, water availability, and crop growing season.
5. Improved understanding of basic soil-water-plant relationships. In-depth research is needed to more fully understand the basic relation-

ships between soil characteristics; tillage practices; water infiltration, evaporation, and storage; and use of soil water by plants under a wide range of climatic conditions and soil types. Such basic knowledge could improve modeling for cropping systems involving surface residues.

REFERENCES

Aase, J.K., and F.H. Siddoway. 1976. Influence of tall wheatgrass wind barriers on soil drying. Agron. J. 68:627–631.

----, and ----. 1980. Stubble height effects on seasonal microclimate, water balance, and plant development of no-till winter wheat. Agric. Meteorol. 21:1–20.

----, ----, and A.L. Black. 1976. Perennial grass barriers for wind erosion control, snow management, and crop production. p. 69–78. *In* R.W. Tinus (ed.) Proc. Symp. on Shelterbelts on the Great Plains, Denver, CO, 20–22 April. Great Plains Agric. Counc. Publ. 78. Great Plains Agric. Council, Lincoln, NE.

Adams, W.E., J.E. Pallas, Jr., and R.N. Dawson. 1970. Tillage methods for corn-sod systems in the Southern Piedmont. Agron. J. 62:646–649.

Allen, R.R., J.T. Musick, and D.A. Dusek. 1980. Limited tillage and energy use with furrow-irrigated grain sorghum. Trans. ASAE 23:346–350.

----, ----, and A.F. Wiese. 1975a. No-till management of furrow irrigated continuous grain sorghum. Texas Agric. Exp. Stn. Prog. Rep. PR-3332 C.

----, ----, and ----. 1976. Limited tillage of furrow irrigated winter wheat. Trans. ASAE 19:234–236, 241.

----, ----, F.O. Wood, and D.A. Dusek. 1975b. No-till seeding of irrigated sorghum double cropped after wheat. Trans. ASAE 18:1109–1113.

Andraski, B.J., T.C. Daniel, B. Lowery, and D.H. Mueller. 1985a. Runoff results from natural and simulated rainfall for four tillage systems. Trans. ASAE 28:1219–1225.

----, D.H. Mueller, and T.C. Daniel. 1985b. Effects of tillage and rainfall simulation date on water and soil losses. Soil Sci. Soc. Am. J. 49:1512–1517.

Army, T.J., A.F. Wiese, and R.J. Hanks. 1961. Effect of tillage and chemical weed control practices on soil moisture losses during the fallow period. Soil Sci. Soc. Am. Proc. 25:410–413.

Baeumer, K., and W.A.P. Bakermans. 1973. Zero-tillage. Adv. Agron. 25:77–123.

Baker, L.O., J.L. Krall, T.S. Aasheim, and T.P. Hartman. 1956. Chemical summer fallow in Montana. Down Earth 11(4):21.

Barnes, O.K., and D.W. Bohmont. 1958. Effects of cropping practices on water intake rates in Northern Great Plains. Wyoming Agric. Exp. Stn. Bull. 358.

----, ----, and F. Rauzi. 1955. Effect of chemical and tillage summer fallow upon water infiltration rates. Agron. J. 47:435–436.

Barnett, A.P. 1987. Fifty years of progress in soil and water conservation research at the Southern Piedmont Conservation Research Center (SPCRC), Watkinsville, GA.

Bauer, A., and H.L. Kucera. 1978. Effect of tillage on some physiochemical properties and on annually cropped spring wheat yields. North Dakota State Univ. Agric. Exp. Stn. Bull. 506.

Beale, O.W., and G.W. Langdale. 1964. The compatibility of corn and coastal bermudagrass as affected by tillage methods. J. Soil Water Conserv. 19:238–240.

----, G.B. Nutt, and T.C. Peele. 1955. The effects of mulch tillage on runoff, erosion, soil properties, and crop yields. Soil Sci. Soc. Am. Proc. 19:244–247.

Bennett, H.H. 1939. Soil conservation. 1st ed. McGraw-Hill Book Co., New York.

Black, A.L., and A. Bauer. 1985. Soil water conservation strategies for Northern Great Plains. p. 76–86. *In* Planning and management of water conservation systems in the Great Plains. Proc. Workshop, Lincoln, NE. 21–25 October. USDA-SCS, Lincoln.

----, and J.F. Power. 1965. Effect of chemical and mechanical fallow methods on moisture storage, wheat, and soil erodibility. Soil Sci. Soc. Am. Proc. 29:465–468.

----, and F.H. Siddoway. 1977. Winter wheat recropping on dryland as affected by stubble height and nitrogen fertilization. Soil Sci. Soc. Am. J. 41:1186–1190.

----, and ----. 1979. Influence of tillage and wheat residue management on soil properties in the Great Plains. J. Soil Water Conserv. 34:220–223.

Borst, H.L., and R. Woodburn. 1942. The effect of mulching and methods of cultivation on runoff and erosion from Muskingum silt loam. Agric. Eng. 23:22–25.

Bristow, K.L., G.S. Campbell, R.I. Papendick, and L.F. Elliott. 1986. Simulation of heat and moisture transfer through a surface residue-soil system. Agric. For. Meteorol. 36:193–214.

Browning, G.M., R.A. Norton, E.V. Collins, and H.A. Wilson. 1945. Tillage practices in relation to soil and water conservation and crop yields in Iowa. Soil Sci. Soc. Am. Proc. 9:241–247.

Bruce, R.R., S.R. Wilkinson, and G.W. Langdale. 1987. Legume effects on soil erosion and productivity. p. 127–138. *In* J.F. Power (ed.) The role of legumes in conservation tillage systems. Soil Conserv. Soc. Am., Ankeny, IA.

Carreker, J.R., J.E. Box, Jr., R.N. Dawson, E.R. Beaty, and H.D. Morris. 1972. No-till corn in fescuegrass. Agron. J. 64:500–503.

Chandler, F.B., and I.C. Mason. 1942. The effect of mulch on soil moisture, soil temperature, and growth of blueberry plants. Am. Soc. Hortic. Sci. Proc. 40:335–337.

Clark, L.E. 1983. Response of cotton to cultural practices. Texas Agric. Exp. Stn. Prog. Rep. PR-4175.

Cochran, V.L., L.F. Elliott, and R.I. Papendick. 1982. Effect of crop residue management and tillage on water use efficiency and yield of winter wheat. Agron. J. 74:929–932.

Dawson, R.C. 1946. Effect of crop residues on soil and moisture conservation under Maryland conditions. Soil Sci. Soc. Am. Proc. 10:425–428.

Duley, F.L., and L.L. Kelly. 1939. Effect of soil type, slope, and surface conditions on intake of water. Univ. of Nebraska Agric. Exp. Stn. Res. Bull. 112.

----, and J.C. Russel. 1939. The use of crop residues for soil and moisture conservation. J. Am. Soc. Agron. 31:703–709.

Free, G.R. 1953. Stubble mulch tillage in New York. Soil Sci. Soc. Am. Proc. 17:165–170.

----, S.N. Fertig, and C.E. Bay. 1963. Zero tillage for corn following sod. Agron. J. 55:207–208.

Gerard, C.J., and D.G. Bordovsky. 1984. Conservation tillage studies in the Rolling Plains. p. 201–216. *In* R.H. Follett (ed.) Conservation tillage. Proc. Great Plains Conserv. Tillage Symp., North Platte, NE. 21–23 August. Great Plains Agric. Counc. Publ. 110. Kansas State Univ., Manhattan.

Gerik, T.J., and J.E. Morrison, Jr. 1984. No-tillage of sorghum on a swelling clay soil. Agron. J. 76:71–76.

----, and ----. 1985. Wheat performance using no-tillage with controlled wheel traffic on a clay soil. Agron. J. 77:115–118.

Greb, B.W. 1979. Reducing drought effects on croplands in the West-Central Great Plains, USDA Info. Bull. 420. U.S. Gov. Print. Office, Washington, DC.

----, and A.L. Black. 1971. Vegetative barriers and artificial fences for managing snow in the Central and Northern Plains. p. 96–111. *In* A.O. Haugen (ed.) Proc. Snow and Ice in Relation to Wildlife and Recreation Symp., Ames, IA. 11–12 February. Iowa Coop. Wildlife Res. Unit, Iowa State Univ., Ames.

----, D.E. Smika, and A.L. Black. 1967. Effect of straw mulch rates on soil water storage during summer fallow in the Great Plains. Soil Sci. Soc. Am. Proc. 31:556–559.

----, ----, N.P. Woodruff, and C.J. Whitfield. 1974. p. 51–85. *In* Summer fallow in the western United States. USDA-ARS Conserv. Res. Rep. 17. U.S. Gov. Print. Office, Washington, DC.

Griffith, D.R., J.V. Mannering, and W.C. Moldenhauer. 1977. Conservation tillage in the eastern Corn Belt. J. Soil Water Conserv. 32:20–28.

Hallsted, A.L., and O.R. Mathews. 1936. Soil moisture and winter wheat with suggestions on abandonment. Kansas Agric. Exp. Stn. Bull. 273.

Harrold, L.L., and W.M. Edwards. 1972. A severe rainstorm test of no-till corn. J. Soil Water Conserv. 27:30.

----, G.B. Triplett, Jr., and W.M. Edwards. 1970. No-tillage corn—Characteristics of the system. Agric. Eng. 51:128–131.

----, ----, and R.E. Youker. 1967. Loess soil and water loss from no-tillage corn. Ohio Rep. 52(2):22–23.

Hays, O.E. 1961. New tillage methods reduce erosion and runoff. J. Soil Water Conserv. 16:172–175.

Horner, G.M., A.G. McCall, and F.G. Bell. 1944. Investigations in erosion control and the reclamation of eroded land at the Palouse Conservation Experiment Station, Pullman, WA, 1931–42. USDA Agric. Tech. Bull. 860. U.S. Gov. Print. Office, Washington, DC.

James, E. 1945. Effect of certain cultural practices on moisture conservation on a Piedmont soil. J. Am. Soc. Agron. 37:945–952.

Jenny, H. 1961. E.W. Hilgard and the birth of modern soil science. Collana Della Revista Agrochimica 3. Simposio Internazionale di Agrochimica, Pisa, Italy.

Johnson, M.D., B. Lowery, and T.C. Daniel. 1984. Soil moisture regimes of three conservation tillage systems. Trans. ASAE 27:1385–1390, 1395.

Johnston, J.C., and W. Sullivan. 1949. Eliminating tillage in citrus soil management. California Agric. Ext. Serv. Circ. 150.

Kidder, E.H., R.S. Stauffer, and C.A. Van Doren. 1943. Effect on infiltration of surface mulches of soybean residues, corn stover, and wheat straw. Agric. Eng. 24:155–159.

Laflen, J.M., J.L. Baker, R.O. Hartwig, W.F. Buchele, and H.P. Johnson. 1978. Soil and water loss from conservation tillage systems. Trans. ASAE 21:881–885.

Langdale, G.W., A.P. Barnett, R.A. Leonard, and W.G. Fleming. 1979. Reduction of soil erosion by the no-till system in the Southern Piedmont. Trans. ASAE 22:82–86, 92.

----, J.E. Box, Jr., C.O. Plank, and W.G. Fleming. 1981. Nitrogen requirements associated with improved conservation tillage for corn production. Commun. Soil Sci. Plant Anal. 12:1133–1149.

----, R.R. Bruce, and A.W. Thomas. 1987. Restoration of eroded Southern Piedmont land in conservation tillage systems. p. 142–143. *In* J.F. Power (ed.) The role of legumes in conservation tillage systems. Soil Conserv. Soc. Am., Ankeny, IA.

----, W.L. Hargrove, and J. Giddens. 1984. Residue management in double-crop conservation tillage systems. Agron. J. 76:689–694.

----, H.F. Perkins, A.P. Barnett, J.C. Reardon, and R.L. Wilson, Jr. 1973. Soil and nutrient runoff losses with in-row, chisel-planted soybeans. J. Soil Water Conserv. 38:297–301.

Larson, W.E., W.C. Burrows, and W.O. Willis. 1960. Soil temperature, soil moisture, and corn growth as influenced by mulches of crop residues. Trans. Int. Congr. Soil Sci. 7th 1:629–637.

Leggett, G.E., R.E. Ramig, L.C. Johnson, and T.W. Massee. 1974. Summer fallow in the northwest. p. 110–135. *In* Summer fallow in the western United States. USDA-ARS Conserv. Res. Rep. 17. U.S. Gov. Print. Office, Washington, DC.

Lillard, J.H., and J.N. Jones, Jr. 1964. Planting and seed-environment problems with corn in killed-sod seedbeds. Trans. ASAE 7:204–206, 208.

Lindstrom, M.J., F.E. Koehler, and R.I. Papendick. 1974. Tillage effects on fallow water storage in the eastern Washington dryland region. Agron. J. 66:312–316.

----, and C.A. Onstad. 1984. Influence of tillage systems on soil physical parameters and infiltration after planting. J. Soil Water Conserv. 39:149–152.

----, W.B. Voorhees, and G.W. Randall. 1981. Long-term effects on interrow runoff and infiltration. Soil Sci. Soc. Am. J. 45:945–948.

Lombard, T.A. 1944. Oil weed control with cultivations versus noncultivation. Calif. Citrogr. 29:212.

Lowdermilk, W.C. 1953. Conquest of the land through seven thousand years. USDA-SCS, Agric. Info. Bull. 99. U.S. Gov. Print. Office, Washington, DC.

Mannering, J.V., and L.D. Meyer. 1961. The effects of different methods of cornstalk residue management on runoff and erosion as evaluated by simulated rainfall. Soil Sci. Soc. Am. Proc. 25:506–510.

----, and ----. 1963. The effects of various rates of surface mulch on infiltration and erosion. Soil Sci. Soc. Am. Proc. 27:84–86.

Massee, T., and H. McKay. 1979. Improving dryland wheat production in eastern Idaho with tillage and cropping methods. Idaho Agric. Exp. Stn. Bull. 581.

Matocha, J.E., and R.C. Bennett. 1984. Tillage systems influence on lint yields and fiber properties of short-season cottons. p. 338. *In* J.M. Brown (ed.) Proc. Beltwide Cotton Prod. Res. Conf., Atlanta, GA. 8–12 January. Natl. Cotton Counc. Am., Memphis, TN.

McCalla, T.M., and T.J. Army. 1961. Stubble mulch farming. Adv. Agron. 13:125–196.

McClellan, W.L. 1970. Sod plantings of corn plots in Pennsylvania. p. 14–17. *In* Randy Schaefer (ed.) Northeastern No-Till Conf., Albany, NY. 1–3 December. Chevron Chem. Co., Richmond, CA.

Meyer, L.D., and J.V. Mannering. 1961. Minimum tillage for corn: Its effect on infiltration and erosion. Agric. Eng. 42:72–75, 86–87.

Middleton, H.E. 1952. Modifying the physical properties of soil. p. 24–41. *In* B.T. Shaw (ed.) Soil physical conditions and plant growth. Academic Press, New York.

Mills, W.C., A.W. Thomas, and G.W. Langdale. 1986. Estimating soil loss probabilities for Southern Piedmont cropping-tillage systems. Trans. ASAE 29:948–955.

Moody, J.E., G.M. Shear, and J.N. Jones, Jr. 1961. Growing corn without tillage. Soil Sci. Soc. Am. Proc. 25:516–517.

Moschler, W.W., G.D. Jones, and G.M. Shear. 1969. Stand and early growth of orchardgrass and red clover seeded after no-tillage corn. Agron. J. 61:475–476.

Musick, J.T., A.F. Wiese, and R.R. Allen. 1977. Management of bed-furrow irrigated soil with limited and no-tillage systems. Trans. ASAE 20:666–672.

Myhre, D.L., and J.O. Sanford. 1972. Soil surface roughness and straw mulch for maximum beneficial use of rainfall by corn on a blackland soil. Soil Sci. 114:373–379.

Naffziger, L.M., and G.M. Horner. 1958. Effect of cropping and tillage practices on runoff and erosion in the Palouse areas of Washington and Idaho. Trans. ASAE 1:34–35.

Nicholaichuk, W., D.M. Gray, H. Steppuhn, and F.B. Dyck. 1986. Snow management practices for trapping snow in a prairie environment. p. 477–499. *In* H. Steppuhn and W. Nicholaichuk (ed.) Proc. Snow Management for Agriculture Symp., Swift Current, Saskatchewan. 9–11 July 1985. Great Plains Agric. Counc. Publ. 120. Swift Current Research Stn., Swift Current, Saskatchewan, Canada.

Onstad, C.A. 1972. Soil and water losses as affected by tillage practices. Trans. ASAE 15:287–289.

Oveson, M.M., and A.P. Appleby. 1971. Influence of tillage management in a stubble mulch fallow-winter wheat rotation with herbicide weed control. Agron. J. 63:19–20.

Papendick, R.I., and D.E. Miller. 1977. Conservation tillage in the Pacific Northwest. J. Soil Water Conserv. 32:49–56.

Parochetti, J.V. 1970. Double cropping. p. 18–22. *In* Randy Schaefer (ed.) Northeastern No-Tillage Conf., Albany, NY. 1–3 December. Chevron Chem. Co., Richmond, CA.

Peele, T.C., G.B. Nutt, and O.W. Beale. 1947. Utilization of plant residues as mulches in the production of corn and oats. Soil Sci. Soc. Am. Proc. 11:356–360.

Phillips, R.E., R.L. Blevins, G.W. Thomas, W.W. Frye, and S.H. Phillips. 1980. No-tillage agriculture. Science 208:1108–1113.

Phillips, W.M. 1954. The use of dalapon and 2,4-D for maintaining summer fallowed land in a weed free condition. p. 44. *In* Res. Rep., 11th North Central Weed Control Conf., Fargo, ND. 7–9 December.

----. 1964. A new technique of controlling weeds in sorghum in a wheat-sorghum-fallow rotation in the Great Plains. Weeds 12:42–44.

----. 1969. Dryland sorghum production and weed control with minimum tillage. Weed Sci. 17:451–454.

Ramig, R.E., R.R. Allmaras, and R.I. Papendick. 1983. Water conservation: Pacific Northwest. *In* H.E. Dregne and W.O. Willis (ed.) Dryland agriculture. Agronomy 23:105–124.

----, and L.G. Ekin. 1978. Soil water storage as influenced by tillage and crop residue management. p. 65–68. *In* Oregon Agric. Exp. Stn. Prog. Rep. SM 78-4.

----, and ----. 1984. Effect of stubble management in a wheat-fallow rotation on water conservation and storage in eastern Oregon. p. 30–33. *In* Oregon Agric. Exp. Stn. Spec. Rep. 713.

Redinger, G.J., G.S. Campbell, K.E. Saxton, and R.I. Papendick. 1984. Infiltration rate of slot mulches: Measurement and numerical simulation. Soil Sci. Soc. Am. J. 48:982-986.

Reicosky, D.C., D.K. Cassel, R.L. Blevins, W.R. Gill, and G.C. Naderman. 1977. Conservation tillage in the southeast. J. Soil Water Conserv. 32:13-19.

Russel, J.C. 1939. The effect of surface cover on soil moisture losses by evaporation. Soil Sci. Soc. Am. Proc. 4:65-70.

Saxton, K.E., D.K. McCool, and R.I. Papendick. 1981. Slot mulch for runoff and erosion control. J. Soil Water Conserv. 36:44-47.

Shear, G.M., and W.W. Moschler. 1969. Continuous corn by the no-tillage and conventional tillage methods: A six-year comparison. Agron. J. 61:524-526.

Smika, D.E. 1976. Seed zone soil water conditions with reduced tillage in the semiarid Central Great Plains. p. 37.1-37.6. *In* Proc. 7th Conf. Int. Soil Tillage Res. Org., Uppsala, Sweden. 13-18 June.

----. 1983. Soil water change as related to position of wheat straw mulch on the soil surface. Soil Sci. Soc. Am. J. 47:988-991.

----. 1985. Strategies and techniques for water conservation with rainfed agriculture in the Central Great Plains. p. 87-96. *In* Planning and management of water conservation systems in the Great Plains states. Proc. Workshop, Lincoln, NE. 21-25 October. USDA-SCS, Lincoln.

----, and P.W. Unger. 1986. Effect of surface residues on soil water storage. Adv. Soil Sci. 5:111-138.

----, and C.J. Whitfield. 1966. Effect of standing wheat stubble on storage of winter precipitation. J. Soil Water Conserv. 21:138-141.

----, and G.A. Wicks. 1968. Soil water storage during fallow in the Central Great Plains as influenced by tillage and herbicide treatments. Soil Sci. Soc. Am. Proc. 32:591-595.

Sojka, R.E., G.W. Langdale, and D.L. Karlen. 1984. Vegetative techniques for reducing water erosion of cropland in the southeastern United States. Adv. Agron. 37:155-181.

Sprague, M.A. 1952. The substitution of chemicals for tillage in pasture renovation. Agron. J. 44:405-409.

Staple, W.J., J.J. Lehane, and A. Wenhardt. 1960. Conservation of soil moisture from fall and winter precipitation. Can. J. Soil Sci. 40:80-88.

Stiegler, J., W. Downs, and F. Hawk. 1984. Lo-till system. p. 197-200. *In* R.H. Follett (ed.) Conservation tillage. Proc. Great Plains Conserv. Tillage Symp., North Platte, NE. 21-23 August. Great Plains Agric. Counc. Publ. 110. Kansas State Univ., Manhattan.

Swan, D.G., M.M. Oveson, and A.P. Appleby. 1974. Chemical and cultural methods for downy brome control and yield of winter wheat. Agron. J. 66:793-795.

Tanaka, D.L. 1985. Chemical and stubble mulch fallow influences on seasonal soil water contents. Soil Sci. Soc. Am. J. 49:728-733.

Taylor, R.E., O.E. Hays, C.E. Bay, and R.M. Nixon. 1964. Corn stover mulch for control of runoff and erosion on land planted to corn after corn. Soil Sci. Soc. Am. Proc. 28:123-125.

Triplett, G.B., Jr., W.H. Johnson, and D.M. Van Doren, Jr. 1963. Performance of two experimental planters for no-tillage corn culture. Agron. J. 55:408-409.

Unger, P.W. 1978. Straw mulch rate effect on soil water storage and sorghum yield. Soil Sci. Soc. Am. J. 42:486-491.

----. 1984. Tillage and residue effects on wheat, sorghum, and sunflower grown in rotation. Soil Sci. Soc. Am. J. 48:885-891.

----, R.R. Allen, and A.F. Wiese. 1971. Tillage and herbicides for surface residue maintenance, weed control, and water conservation. J. Soil Water Conserv. 26:147-150.

----, C.J. Gerard, J.E. Matocha, F.M. Hons, D.G. Bordovsky, and C.W. Wendt. 1987. Water management with conservation tillage. p. 16-22. *In* T.J. Gerik and B.L. Harris (ed.) Conservation tillage: Today and tomorrow. Proc. South. Region No-Till Conf., College Station, TX. 1-2 July. Texas Agric. Exp. Stn. MP-1636.

----, J.L. Steiner, and O.R. Jones. 1986. Response of conservation tillage sorghum to growing season precipitation. Soil Tillage Res. 7:291-300.

----, and A.F. Wiese. 1979. Managing irrigated winter wheat residues for water storage and subsequent dryland grain sorghum production. Soil Sci. Soc.Am. J. 43:582–588.

Van Doren, D.M., Jr., and G.B. Triplett, Jr. 1969. Mechanism of corn (*Zea mays* L.) response to cropping practices without tillage. Ohio Agric. Res. Dev. Ctr. Res. Circ. 169.

Wicks, G.A., and D.E. Smika. 1973. Chemical fallow in a winter wheat-fallow rotation. J. Weed Sci. Soc. Am. 21:97–102.

Wiese, A.F. 1966. Chemical fallow still costs too much. Crops Soils 18(8):15.

----, and T.J. Army. 1958. Effect of tillage and chemical weed control practices on soil moisture storage and losses. Agron. J. 50:465–468.

----, and T.J. Army. 1960. Effect of chemical fallow on soil moisture storage. Agron. J. 52:612–613.

----, J.J. Bond, and T.J. Army. 1960. Chemical fallow in the Southern Great Plains. Weeds 8:284–290.

----, and D.W. Staniforth. 1973. Weed control in conservation tillage. p. 108–114. *In* Max Schnepf (ed.) Conservation tillage. Proc. Natl. Conf., Des Moines, IA. 28–30 March. Soil Conserv. Soc. Am., Ankeny, IA.

Willis, W.O., and A.B. Frank. 1975. Water conservation by snow management in North Dakota. p. 155–162. *In* Proc. Snow Management on the Great Plains, Bismark, ND. 29 July. Great Plains Agric. Counc. Publ. 73. Univ. of Nebraska, Lincoln.

Zingg, A.W., and C.J. Whitfield. 1957. A summary of research experience with stubble mulch farming in the western states. USDA Tech. Bull. 1166. U.S. Gov. Print. Office, Washington, DC.

6 Role of Crop Residue Management in Nitrogen Cycling and Use[1]

J. F. Power and J. W. Doran
USDA-ARS
University of Nebraska-Lincoln
Lincoln, Nebraska

Reduced or no tillage is presently used on 25 to 40 million ha in the USA. This figure is projected to double within the next decade or two (Magleby et al., 1985). Fertilizer requirements and nutrient availability change with tillage and residue management, dictating the need for a more thorough and basic understanding of the effects of management practices on the cycling, transformations, and availability of the nutrients needed for crop production. Such information is needed as a base for tillage management and cropping strategies that protect the environment while maintaining economic viability. Choice of cropping and management practices used with a given soil in a given crop production enterprise play a major role in determining the efficiency with which water and N resources are utilized. Practices best suited for one soil-crop combination may be inefficient for another combination. In this chapter, the principles controlling the interactions between soil, cropping system, and crop-residue-management practices are outlined and discussed, pointing out how the effects of these production variables alter the cycling and use of N from all sources.

The severe soil erosion that occurred in North America in the 1930s emphasized the need for development of tillage practices that would control erosion and maintain productivity. At about this time, Duley and Russel (1939) began publications of their research on stubble-mulch farming, illustrating the feasibility of leaving some crop residues on the soil surface. Although it has long been recognized that crop residues may affect nutrient cycling and transformations, effects of residue-management practices on nutrient availability have been extensively evaluated only in more recent years. Results of early research often indicated that stubble-mulch tillage resulted in less N available for crop growth than occurred with bare tillage (Winterlin et al., 1958). With more recent widespread use of herbicides and introduc-

[1] Contribution from the USDA-ARS, in cooperation with the Agric. Res. Div., Univ. of Nebraska-Lincoln, Lincoln, NE 68583.

Cropping Strategies for Efficient Use of Water and Nitrogen, Special Publication no. 51.

tion of no-till production methods, the problem of immobilization of N in crop residues and other organic forms became more acute. Thus, a number of detailed summaries on N cycling as affected by tillage and crop-residue management have been reported more recently (Campbell et al., 1976; Doran, 1980; Fox and Bandel, 1986; Gilliam and Hoyt, 1987).

TILLAGE/RESIDUE MANAGEMENT

Tillage has two effects on the soil environment that may act independently or in unison. Tillage alters soil geometry by disturbing the soil, thereby affecting pore size and frequency, size distribution of aggregates, soil color, bulk density, and related properties. Soil tillage also affects the placement and distribution of crop residues, either leaving them on the soil surface or incorporating them to some degree into the soil. In practice, with the use of tillage, the soil is physically disturbed simultaneously with residue placement because it is through tillage that crop residues are managed. In the following sections, we attempt to reduce the complexity of tillage- and residue-management interactions on N cycling by discussing them separately to the extent possible.

The effects of soil disturbance resulting from tillage on organic matter and soil porosity and their subsequent effects on water content are illustrated by data in Table 6–1 from work at several research sites across the USA (Mielke et al., 1986). Compared to no-tillage, plowing generally increased soil porosity and decreased bulk density within the tilled zone. Increased porosity in the plowed soil compared to that with no-tillage results in shallower penetration of a given quantity of water added by either precipitation or irrigation. For semiarid wheat regions such as western Nebraska, much of the growing-season precipitation is received in showers of 15 mm or less. Because depth of water penetration is less in the more porous plowed soil, the potential for loss through evaporation is greater, when compared to no-tilled soil. This fact partially accounts for the more favorable water conservation efficiency usually found with no-till.

During the period when bulk density is reduced following tillage, total porosity is also greater in tilled than in no-tilled soil. The increased total pore volume, coupled with reduced water storage in tilled soil, results in a large decrease in water-filled pore space following tillage. This change increases the rate of O_2 diffusion through soil, creating a more aerobic and arid soil environment than occurs in no-tilled soil. Bulk density is usually greater and porosity is less in cultivated soils than in corresponding soils under grass sod (Broder et al., 1984).

Because of the greater porosity of plowed compared to no-till soil, hydraulic conductivity in surface soils is often greater in plowed soil (Table 6–1). Although not shown in these data, hydraulic conductivity beneath the surface layer is often greater for no-till than for plowed soil (Mielke et al., 1984). This may result from greater continuity of macropores, often resulting from undisturbed earthworm and old root channels being maintained almost

Table 6-1. Surface soil physical properties (0-75 mm) as affected by tillage method at seven experimental sites in the USA. After Mielke et al. (1986).

Crop, soil location	Tillage management	Soil property						
		Bulk density	Organic carbon	Water content	Porosity	Water-filled pore space	Air permeability	Hydraulic conductivity
		g cm^{-3}	%, w/v	——	v/v	——	pm^2	mm h^{-1}
Corn								
Blount sil†	No-till	1.42*	1.84*	0.29*	0.46	0.62*	1	<1
Illinois	Plow	1.30	1.24	0.17	0.51*	0.34	3*	<1
Maury sil	No-till	1.26	2.82*	0.34*	0.52	0.65*	2	4
Kentucky	Plow	1.26	1.61	0.24	0.52	0.47	6*	7
Nicollet cl	No-till	1.30*	3.81*	0.34*	0.51	0.67*	2	1
Minnesota	Plow	1.17	3.20	0.25	0.56*	0.45	11*	9*
Webster cl	No-till	0.96*	3.92	0.34	0.64	0.53	<1	<1
Minnesota	Plow	0.88	3.50	0.32	0.67*	0.48	24*	26*
Crete-Butler sicl	No-till	1.30*	2.51*	0.25	0.51	0.50*	<1	<1
Nebraska	Plow	1.04	1.73	0.21	0.61*	0.35	22*	10*
Wheat								
Alliance sil	No-till	1.29	1.32*	0.30*	0.51	0.58*	5	4
Nebraska	Plow	1.25	1.00	0.28	0.53	0.52	11	18*
Duroc l	No-till	1.02	2.24*	0.21*	0.62	0.34*	5	40
Nebraska	Plow	1.18*	1.70	0.18	0.54*	0.33	21*	148

* Tillage management means within location differ significantly at $P < 0.05$.
† Sil = silt loam, cl = clay loam, l = loam.

to the soil surface with no-till (Douglas et al., 1980). While water transport through these macropores in no-till soil may be relatively rapid, this water comes in little contact with the microbial microsites predominantly found in smaller diameter pores, so probably results in little leaching of NO_3-N to deeper depths.

The changes in water content and aeration that result from tillage are reflected in soil-temperature regimes. The greater heat sink of the additional water in the no-tilled soil, along with slower rates of gaseous exchange, result in less variation in diurnal temperatures and a slower rate of temperature change. This, in addition to the insulating effects of crop residues when left on the soil surface, reduces mean daily soil temperatures. These effects of surface residue placement were documented by McCalla and Duley (1946) and have since been frequently verified. Normally, mean daily temperatures are reduced 2 to 4°C by surface residue placement. Reductions in soil temperature, however, and consequent effects on biological activity are dependent on climate, soils, and time of year (Doran and Smith, 1987).

EFFECTS OF CROP RESIDUES ON SOIL PROPERTIES

Crop residue management may alter many soil properties—physical, chemical, and biological. Some of these effects were identified in the previous

Table 6-2. Effect of surface application of wheat straw (6 yr) on soil properties (Black, 1973).

Residue rate	Depth	Organic matter	Total N	Aggregates over 8.4 mm	Bulk density	Mineral-izable NO_3-N	$NaHCO_3^-$ soluble P
kg ha^{-1} yr^{-1}	mm	g kg^{-1}			Mg m^{-3}	mg kg^{-1}	
0	0-75	17.9	0.89	500	1.38	18.2	3.9
	75-150	13.3	0.72	500	1.63	18.2	3.9
1680	0-75	19.9	0.97	540	1.31	21.7	4.4
	75-150	14.0	0.74	540	1.58	21.7	4.4
3360	0-75	21.1	0.96	618	1.29	21.1	4.3
	75-150	15.0	0.83	618	1.56	21.1	4.3
6730	0-75	22.0	1.02	725	1.27	23.3	4.9
	75-150	17.1	0.87	725	1.55	23.3	4.9

section (Table 6-1). Changes in many of these soil properties affect the microbiological environment within the soil, thereby altering nutrient transformations and efficiency of use. Particularly important in characterizing the soil environment are the water, aeration, temperature, and substrate (nutrient) regimes established by crop residue management.

A number of field experiments have reported on the effects of crop residue management on a number of soil properties. Larson et al. (1972) found that cornstalks added at rates of at least 6 Mg ha^{-1} could maintain or increase organic carbon and nitrogen contents of a Marshall silty clay loam (fine-silty, mixed, mesic Typic Hapludoll) in Iowa. Black (1973) in Montana showed that, in a 6-yr period, increasing the quantity of wheat (*Triticum aestivum* L.) straw applied to the soil surface increased the amount of organic carbon and nitrogen, mineralizable N, $NaHCO_3^-$ soluble P, and aggregate size in the surface soil (Table 6-2). Biederbeck et al. (1984) reported somewhat similar results from Saskatchewan for various cropping systems.

The effects of crop-residue management on soil properties affect almost all parameters of the soil environment. The above publications, as well as many others, document these effects in regard to the size of a number of chemical pools (especially for C and N), water infiltration and movement, temperature, and aeration. The interactions among these factors define the habitats being established for biological activity in the soil, but these interactions are often difficult to quantify under ambient conditions.

Degree of aggregation is one expression of the interactions between physical, chemical, and biological forces involved in soils. Tisdale and Oades (1982) and others have concluded that the development and activity of fungal hyphae play major roles in soil aggregation. Biederbeck et al. (1984) also indicated that crop residue management affects microbial activity and biomass as well as soil aggregation. Data presented in Table 6-2 from Black (1973) likewise show that crop residues affect aggregation. Campbell et al. (1986) found that manure application had similar effects of stabilizing soil aggregation, probably through the production of humic colloids. These and other publications (Campbell, 1978) suggest that, by leaving organic residues on the soil surface, microbial activity and growth are stimulated, resulting not

Table 6-3. Tillage effects on microbial populations, microbial biomass, and potentially mineralizable N for six research sites across the USA. After Linn and Doran (1984a) and Doran (1987).

	Ratio—No-till/plow		
Soil variable	0-75 mm	75-150 mm	150-300 mm
Total aerobes	1.35	0.66	0.82
Fungi	1.35	0.55	0.69
Aerobic bacteria	1.41	0.68	0.76
NH_4 oxidizers†	1.25	0.55	--
NO_2 oxidizers†	1.58	0.75	--
Facultative anaerobes	1.31	0.96	0.94
Obligate anaerobes	1.27	1.05	1.01
Denitrifiers†	7.31	1.77	--
PMN‡	1.37	0.98	0.93
Microbial biomass	1.54	0.98	1.00

† After Doran (1980) for seven sites. ‡ PMN = Potentially mineralizable N.

only in greater pools of labile N, but also improved soil physical conditions that maintain porosity and promote aeration and water regimes favorable for biological activity.

BIOLOGICAL RESPONSE TO CHANGES IN SOIL ENVIRONMENT

Biological responses to changes in soil environment resulting from tillage and crop residue placement are discussed relative to (i) responses of soil microorganisms and (ii) responses of crop plants. Alteration of the soil environment through residue management changes the characteristics and distribution of biological niches in the soil. Various groups of soil microorganisms proliferate and become active in response to the niches created. As indicated in the previous section, removal of surface residues through tillage, burning, or other means indirectly reduces soil water content, increases soil temperature, and increases air-filled pore space. The more aerobic and arid soil conditions created by such action enhance the growth and activity of aerobic microorganisms relative to those of anaerobes. Among the bacteria, aerobes are favored over obligate anaerobes, and facultative anaerobes would more often be metabolizing aerobically. These changes in microbial populations and activity would encourage mineralization and subsequent nitrification of soil organic nitrogen and decrease denitrification loss of NO_3-N. Conversely, maintenance of surface residues with no-tillage usually increases soil water content, decreases air-filled pore space, and decreases O_2 diffusion. Consequently, the potential for denitrification may be enhanced, particularly in humid climates. In poorly drained soils, denitrification under no-till may be great, resulting in inefficient use of N.

Doran (1980) documented the changes in microbial populations that occur as a result of different tillage practices (Table 6-3). Effects of tillage and crop residue management are especially pronounced in the upper 75 mm of soil. At this depth, populations of all microorganisms were greater for no-tillage than for plowing, with greatest effects on anaerobic organisms. At the 75- to 150-mm and 150- to 300-mm depths, however, populations for all aerobic organisms were reduced by no-tillage. Likewise, potentially mineralizable N and microbial biomass levels in the surface 75 mm were greater for no-tillage, but not at the deeper depths. For these six sites, Doran (1987) found that potentially mineralizable N increased as mean annual precipitation increased, but with a greater response to increased precipitation for no-tilled (surface residues) than for plowed soil.

A number of scientists have studied how microbial populations and aerobic and anaerobic activity fluctuate with soil porosity and water content, and the influence of these factors on production of N_2O and CO_2. Shortly after irrigation or rainfall, greater soil water content and reduced air-filled porosity in no-tilled compared to plowed soils are major causes of greater denitrification rates and gaseous N losses with no-tillage (Aulakh et al., 1982; Linn and Doran, 1984a, 1984b; Rice and Smith, 1982).

In addition to physical factors, the soil chemical environment also profoundly affects microbial populations and activity. As indicated by Paul and Juma (1981), the amount and location of soluble organic carbon substrates are of major importance in providing energy for the metabolism of microorganisms. Location of C-rich crop residues (often with a relatively wide C/N ratio) is controlled to a large degree by tillage practices. As a result, with no-till systems, the C-rich residues are left on the surface so microbial populations in the surface soil are increased (Table 6–3). The stratification of crop residues and concentration of soil organic matter near the soil surface with reduced or no-tillage management are paralleled by greater soil microbial biomass levels, which have been related to greater immobilization of fertilizer N as compared with plowing or shallow tillage (Rice et al., 1986; Carter and Rennie, 1987). In comparing surface and buried crop residues, qualitative differences in decomposing microbial populations have been identified as factors stimulating N immobilization and organic-matter conservation in reduced and no-tillage management systems (Holland and Coleman, 1987). Slower decomposition and the concentration of crop residues and organic matter near the soil surface increase the diversity of microorganisms and fauna, which can result in greater recycling not only of N within the soil ecosystem but also creates less potential for loss of plant-available N due to leaching compared with conventional moldboard plow tillage (House et al., 1984). Consequently, no-tillage systems may help maintain groundwater quality.

The distribution and activity of plant roots with depth in soil, factors influencing N cycling and availability, are also influenced by tillage and crop residue management. Plant root growth in response to water, temperature, soil resistance, nutrient deficiencies, and other factors has been studied frequently; however, we can only make a few generalized statements relating to the ecology of the species studied. Roots of most crop plants appear to be most active in soils where water content is near field capacity, temperatures are in the 20 to 30 °C range, nutrients are readily available in the soil solution, toxic elements or compounds are absent, and soil aeration is adequate.

Table 6-4. Mean density of wheat roots in upper 0.9 m of soil for three fallow tillage practices at three sampling dates during 1978 (Wilhelm et al., 1982).

Tillage practice	Sampling date			
	28 March	2 May	8 June	$\bar{x}$
	g roots/m³			
No-till	12	60	62	46
Subtill	9	25	44	26
Plow	5	36	59	33

Rice (*Oryza sativa* L.) is an exception because of structural adaptations that enable roots of these plants to grow in saturated soil.

Plant root growth responses to crop residue placement were documented by Wilhelm et al. (1982) for winter wheat (*Triticum aestivum* L.) produced under bare, stubble-mulch, or no-till fallow (Table 6-4). The greater root growth for wheat on no-till fallowed soil may have resulted from more favorable temperature and water regimes during the growing season than for wheat on bare fallow. In another study, Newell and Wilhelm (1987) found that crop residues on the surface also enhanced growth of roots for both irrigated and dryland corn (*Zea mays* L.). However, detailed studies under controlled conditions need to be conducted to measure root growth, distribution, and activity as influenced by controlled changes in the soil physical environment. Grable and Siemer (1968) indicated that root activity may also be explained on the basis of air-filled (or water-filled) pore space in the soil, but additional research is needed on this subject.

In addition to the effects of soil environment on plant root growth and activity, conversely plant roots are a major factor in controlling soil environment. Physically, plant roots penetrate compacted soils and aid in reducing bulk density of soils. Likewise, old plant root channels often serve as macropores and are important in the movement of water and solutes (Douglas et al., 1980). Plant roots also serve to bind and aggregate some soils (Campbell, 1978). These direct physical effects may have a major effect on water and aeration regimes in a soil.

Plant roots also exert direct chemical and biological influences on the soil environment. They are a major source of organic carbon needed for many microbial processes, and roots are particularly effective in this role because of their intimate contact with soil pores (Campbell, 1978). Plant roots also excrete CO_2 and often various organic compounds that may have various effects on microbial activity. Normally, microbial populations in the rhizosphere soil are severalfold greater in numbers than those in nonrhizosphere soil.

NITROGEN TRANSFORMATIONS AND CROP UPTAKE

As explained in the preceding section, the N transformations that occur in a soil are the net result of the kind, quantity, and activity of soil micro-

Table 6-5. Relative quantity, turnover rate, and availability of various soil organic nitrogen components of prairie soils (Campbell et al., 1976).

Fraction	Total soil N	Half-life	Mineralizable N pool
	%	yr	%
Plant residues			
Living	6	--	--
Dead	4	0.3	--
Microbial biomass	5	1.2	68.3
Labile organic nitrogen	65	36.0	31.3
Resistant N	20	990.0	0.4

organisms present—which, in turn, depend largely on the soil physical and chemical environment created by crop residue placement or other soil management practices. Numerous studies have shown that soil organic nitrogen levels are usually enhanced by returning increased rates of crop residues to the soil (Table 6-2). Some of this increase in soil organic nitrogen is in the more readily labile forms. Lynch and Panting (1980), Doran (1987), and others have shown that an appreciable part of the increase in labile soil organic nitrogen in surface soils with reduced tillage is accounted for by the soil microbial biomass. Much microbial N potentially can be mineralized during the course of a growing season (Paul and Juma, 1981) and become available for plant uptake. The rate at which this occurs, however, depends greatly on soil environmental conditions and their fluctuation during the growing season. Some of the N contained in the crop residues may be utilized by new microbial growth and contribute to the mineralizable N pool (Fribourg and Bartholomew, 1956; Power et al., 1986). The relative quantity and availability of several components of the soil organic nitrogen pool are given in Table 6-5 (Campbell et al., 1976).

Stanford (1982) and co-workers postulated that the availability and uptake of mineralizable N depended upon the water and temperature regimes of the soil during the growing season. The validity of this concept was verified in field experiments by Smith et al. (1977). Increased use of surface residues would normally increase soil water content and decrease soil temperature. Depending on the ambient levels of these two factors, then, use of surface residues could have a variable effect upon the amount of soil N mineralized under field conditions during the growing season—a negative effect in cold or poorly drained soils and a positive effect in warm or well-drained soils.

The chemical composition of the crop residues present is important in rate of N cycling that occurs. Greater microbial biomass resulting from surface placement of crop residues with a wide C/N ratio results in a greater potential for immobilization of surface-applied fertilizer N than would occur if crop residues were incorporated. Microbial growth requirements for N are greatest during periods of optimum water and temperature regimes, so microorganisms are in direct competition with growing plants for available N. The N immobilized in microbial biomass is then released to the environment during periods less favorable for biological activity because the supply of available C and demand for N are less at such times (Smith and Blevins, 1987). The wider C/N ratio of the residues, the longer this process requires.

Table 6-6. Recovery of ^{15}N by wheat and in soil as affected by surface crop residues (Myers and Paul, 1971).

		^{15}N recovered in:		
^{15}N applied	Surface residues	Plant	Soil	Total
kg/ha		% ^{15}N applied		
56	Without	29.4	25.4	64.8
	With	21.0	43.4	64.4
112	Without	22.1	44.1	66.2
	With	20.7	44.6	65.3

Because the soluble C used by soil microorganisms is usually derived from crop residues, potential for temporary N immobilization in microbial biomass is great near the surface of no-tilled soils, particularly for straw or stover with a wide C/N ratio. Consequently, efficiency of recovery of surface-applied fertilizer N is often less in soils covered by surface residues than in bare soils (Table 6-6). One obvious solution to this problem would be to concentrate the fertilizer below the C-enriched surface soil that results from surface placement of residues (Doran and Smith, 1987; Mengel et al., 1982). This practice can greatly enhance fertilizer N-use efficiency in no-till soils.

The cooler and more moist soil environment created by surface placement of residues can promote N loss through both denitrification and leaching. Denitrification occurs more rapidly in the field, especially where surface residues are maintained on poorly drained soils. Bedding of the soil or other precautions need to be taken to enhance aeration and reduce these losses. However, even in well-drained soils in semiarid environments, field evidence of denitrification (Aulakh et al., 1982) and of leaching (Herron et al., 1968) have been documented. Isotopic tracer studies frequently show that, in such soils, about 20% of the fertilized N is lost by these two mechanisms combined.

Crop residue management affects the utilization of N not only from crop residues, but also from soil organic matter, and applied fertilizer N. Using N isotopes, Power et al. (1986) found that by increasing the amount of crop residues on the surface of no-tilled soils from 0 to 150% of that produced by the previous crop (up to 6–8 Mg ha^{-1}), total N uptake by soybean [*Glycine max* (L.) Merr.] or corn increased twofold (Table 6-7). Likewise, with increased residue rate, mineralization and uptake of indigenous soil N by corn was increased from 73 to 124 kg ha^{-1}, while that for soybean increased from 84 to 106 kg ha^{-1} (for soybean, biologically fixed N was included with indigenous soil N uptake). Small amounts of N in corn residues were mineralized and used by the next crop, but as much as 63 kg of N ha^{-1} in soybean residues were used. Crop residue rate had less effect upon recovery of fertilizer N either in the year of application or from residual effects from fertilization the previous year.

In this same study, Wilhelm et al. (1986) showed that each megagram of surface residue increased soil water storage 6 to 8 mm, resulting in corn and soybean grain yield increases of 120 and 90 kg ha^{-1}, respectively, per

Table 6-7. Sources of N taken up by soybean and corn by harvest in 1981 (Power et al., 1986).

Residue rate†	Nitrogen source: Crop residues	Residual fertilizer	Current fertilizer	Native soil N	Total
	kg N ha^{-1}				
Soybean					
0.0	0.0	1.8	13.8	84.2‡	99.8
0.5	1.0	1.6	20.6	124.0‡	147.2
1.0	38.2	6.7	16.1	115.9‡	176.9
1.5	63.2	5.7	20.0	105.5‡	194.4
Corn					
0.0	0.0	4.8	3.7	72.8	81.3
0.5	0.0	5.8	7.3	96.7	109.8
1.0	1.8	5.8	7.1	114.4	129.1
1.5	0.0	5.7	10.6	124.5	140.8

† As fraction of that produced by previous crop. ‡ Includes biologically fixed N.

megagram residues. These results confirm earlier research by Black (1973) and others that increased quantities of crop residues on the soil surface increase soil water storage and reduce surface soil temperatures, resulting in potential for increased grain yield. Most of the additional N needed to support the increased yield came from increased mineralization of the indigenous soil organic nitrogen, and some also came from the crop residues themselves. It is probable that the extra water conserved in residue-covered soil, along with more favorable soil temperature regimes, not only enhanced plant growth, but also stimulated activity of soil organisms involved in the mineralization of indigenous soil N.

Crop residues ordinarily originate from the harvest remnants of the previous crop. A special situation is created by the planting of cover crops, increasing the amount of crop residues at planting time for the next grain crop. For continuous corn, winter annuals such as rye (*Secale cereale* L.), hairy vetch (*Vicia villosa* Roth), or crimson clover (*Trifolium incarnatum* L.) are frequently used for winter cover. The use of legumes for this purpose has potential of also adding biologically fixed N to the N cycle. Cover crops also offer soil erosion protection and utilize residual soil nitrates, thereby reducing potential for denitrification or leaching losses. Biederbeck et al. (1984) found that rye winter cover utilized residual NO_3^--N, increasing mineralizable N- and N-use efficiency by the following grain crop.

Use of legume cover crops frequently reduces fertilizer-N requirements of the next grain crop by 50 to 120 kg ha^{-1} (Power, 1987). Also, N immobilized in legume biomass is mineralized during the next growing season, providing the following grain crop with a slow-release source of available N. Use of hairy vetch as a winter cover crop with ridge-till continuous corn offers promise of greatly reducing both fertilizer and herbicide input costs.

CONCLUSION

Crop residue management alters the soil environment, thereby influencing microbiological populations and activity in the soil and subsequent N transformations. It is through this chain of events that crop residue management affects the efficiency with which fertilizers, water, and other resources are used in a cropping system. Because so many variables are affected by crop residue management, it is difficult to predict accurately the outcome of a particular management practice over the entire range of crops, soils, and climates encountered.

In spite of the complexities involved, however, a few generalized conclusions regarding crop residue management are possible. First, leaving crop residues on the soil surface reduces evaporation rate, thereby maintaining soil water regimes that are more favorable for microbiological activity over longer periods of time. A consequence of this phonomenon is enhanced size and duration of the soil microbial biomass and greater quantities of N temporarily immobilized in this biomass. Research at several locations suggests that, under certain environmental conditions, the N immobilized in microbial biomass is a primary source of the N mineralized and made available to a crop during the growing season. Thus, the enhanced microbial biomass in the surface of residue-covered soils would contribute toward greater quantities and more controlled rates of N mineralization. Especially for summer crops such as corn, the controlled rate of N release would contribute to more efficient uptake and utilization of N in the cropping system. Within limits, the greater the quantity of residues involved, the greater would be the effect upon N mineralization and efficiency of use.

There are several other mechanisms by which crop residue management may affect N cycling and efficiency of use. An example would be reduced losses of soil organic nitrogen through erosion, thereby maintaining organic nitrogen pool sizes and turnover rates. Another example would involve effects of cropping systems, with rotations containing legumes providing crop residues with narrower C/N ratios than found with grain crops. This change in C/N ratio would again affect N pool sizes and turnover rates, thereby affecting N-use efficiency. In a similar manner, most other soil management practices could affect these mechanisms to some degree.

Given the number of variables involved and the complexities of interactions between soils, crops, and climate, an accurate prediction of the effects of crop-residue-management practices on N cycling and efficient use is possible only through the development of realistic process-based computer simulation models. With such models, a producer on a given soil within a given climate could determine the consequences of alternative residue management practices and how such practices would affect N availability and use, crop production, environmental quality, and net profit. Much additional research is needed, however, before we understand and can accurately quantify the various processes involved.

REFERENCES

Aulakh, M.S., D.A. Rennie, and E.A. Paul. 1982. Gaseous losses from cropped and summerfallowed fields. Can. J. Soil Sci. 62:187–196.

Biederbeck, V.O., C.A. Campbell, and R.P. Zentner. 1984. Effect of crop rotation and fertilization on some biological properties of a loam soil in southwestern Saskatchewan. Can. J. Soil Sci. 64:355–367.

Black, A.L. 1973. Soil property changes associated with crop residue management in a wheat-fallow rotation. Soil Sci. Soc. Am. Proc. 37:943–946.

Broder, M.W., J.W. Doran, G.A. Peterson, and C.R. Fenster. 1984. Fallow tillage influence on spring populations of soil nitrifiers, denitrifiers, and available nitrogen. Soil Sci. Soc. Am. J. 48:1060–1067.

Campbell, C.A. 1978. Soil organic carbon, nitrogen and fertility. p. 173–271. *In* M. Schnitzer and S.U. Khan (ed.) Soil organic matter. Developments in soil science 8. Elsevier Sci. Publ. Co., Amsterdam.

----, E.A. Paul, and W.B. McGill. 1976. Effect of cultivation and cropping on the amounts and forms of soil N. p. 9–101. *In* Proc. Western Canada Nitrogen Symp. Alberta Agric., Edmonton, Alberta.

----, M. Schnitzer, J.W.B. Stewart, V.O. Biederbeck, and F. Selles. 1986. Effect of manure and P fertilizer on properties of a Black Chernozem in southern Saskatchewan. Can. J. Soil Sci. 66:601–613.

Carter, M.R., and D.A. Rennie. 1987. Effects of tillage on deposition and utilization of ^{15}N residual fertilizer. Soil Tillage Res. 9:33–43.

Doran, J.W. 1980. Soil microbial and biochemical changes associated with reduced tillage. Soil Sci. Soc. Am. J. 44:765–771.

----. 1987. Microbial biomass and mineralizable nitrogen distributions in no-tillage and plowed soils. Biol. Fert. Soils 5:68–78.

----, and M.S. Smith. 1987. Organic matter management and utilization of soil and fertility nutrients. p. 51–70. *In* R.F. Follett et al. (ed.) Soil fertility and organic matter as critical components of production systems. Spec. Publ. 19. ASA, SSSA, and CSSA, Madison, WI.

Douglas, J.T., M.J. Goss, and D. Hill. 1980. Measurement of pore characteristics in a clay soil under ploughing and direct drilling, including use of a radioactive tracer (144-Ce) technique. Soil Tillage Res. 1:11–18.

Duley, F.L., and J.C. Russel. 1939. The use of crop residues for soil and moisture conservation. J. Am. Soc. Agron. 31:703–709.

Fox, R.H., and V.A. Bandel. 1986. Nitrogen utilization with no-tillage. p. 117–148. *In* M.A. Sprague and G.B. Triplett (ed.) No-tillage and surface-tillage agriculture: The tillage revolution. John Wiley and Sons, New York.

Fribourg, H.A., and W.V. Bartholomew. 1956. Availability of nitrogen from crop residues during the first and second seasons after application. Soil Sci. Soc. Am. Proc. 20:505–508.

Gilliam, J.W., and G.D. Hoyt. 1987. Effect of conservation tillage on fate and transport of nitrogen. p. 217–240. *In* T.J. Logan et al. (ed.) Effects of conservation tillage on groundwater quality: Nitrates and pesticides. Lewis Publ., Chelsea, MI.

Grable, A.R., and E.G. Siemer. 1968. Effect of bulk density, aggregate size, and soil water suction on oxygen diffusion, redox potentials, and elongation of corn roots. Soil Sci. Soc. Am. Proc. 32:180–186.

Herron, G.M., G.L. Terman, A.F. Dreier, and R.A. Olson. 1968. Residual nitrate nitrogen in fertilized deep loess-derived soils. Agron. J. 60:477–482.

Holland, E.A., and D.C. Coleman. 1987. Litter placement effects on microbial and organic matter dynamics in an agroecosystem. Ecology 68:425–433.

House, G.J., B.J. Stinner, D.A. Crossley, Jr., E.P. Odum, and G.W. Langdale. 1984. Nitrogen cycling in conventional and no-tillage agroecosystems in the Southern Piedmont. J. Soil Water Conserv. 39:194–200.

Larson, W.E., C.E. Clapp, W.H. Pierce, and Y.B. Morachan. 1972. Effect of increasing amounts of residues on continuous corn: Organic carbon, nitrogen, phosphorus, and sulfur. Agron. J. 64:204–208.

Linn, D.M., and J.W. Doran. 1984a. Aerobic and anaerobic microbial populations in no-till and plowed soils. Soil Sci. Soc. Am. J. 48:794–799.

----, and ----. 1984b. Effect of water-filled pore space or CO_2 and N_2O production in tilled and nontilled soils. Soil Sci. Soc. Am. J. 48:1167–1172.

Lynch, J.M., and L.M. Panting. 1980. Cultivation and the soil biomass. Soil Biol. Biochem. 12:29–33.

Magleby, R., D. Gadsby, D. Colacicco, and J. Thigpen. 1985. Trends in conservation tillage use. J. Soil Water Conserv. 40:274–276.

McCalla, T.M., and F.L. Duley. 1946. Effect of crop residues on soil temperature. J. Am. Soc. Agron. 38:75–89.

Mengel, D.B., D.W. Nelson, and D.M. Huber. 1982. Placement of fertilizers for no-till and conventional till corn. Agron. J. 74:515–518.

Mielke, L.N., J.W. Doran, and K.A. Richards. 1986. Physical environment near the surface of plowed and no-tilled surface soils. Soil Tillage Res. 7:355–366.

----, W.W. Wilhelm, K.A. Richards, and C.R. Fenster. 1984. Soil physical charactersitics of reduced tillage in a wheat-fallow system. Trans. ASAE 27:1724–1728.

Myers, R.J.K., and E.A. Paul. 1971. Plant uptake and immobilization of ^{15}N-labelled ammonium nitrate in a field experiment with wheat. p. 55–64. *In* Nitrogen-15 in soil-plant studies. IAEA, Vienna.

Newell, R.L., and W.W. Wilhelm. 1987. Conservation tillage and irrigation effects on corn root development. Agron. J. 79:160–165.

Paul, E.A., and N.G. Juma. 1981. Mineralization and immobilization of soil nitrogen by microorganisms. *In* F.E. Clark and T. Rosswall (ed.) Terrestrial nitrogen cycles. Ecol. Bull. (Stockholm) 33:179–195.

Power, J.F. (ed.) 1987. The role of legumes in conservation tillage systems. Soil Conserv. Soc. Am., Ankeny, IA.

----, W.W. Wilhelm, and J.W. Doran. 1986. Crop residue effects on soil environment and dryland maize and soya bean production. Soil Tillage Res. 8:101–111.

Rice, C.W., and M.S. Smith. 1982. Denitrification in no-till and plowed soils. Soil Sci. Soc. Am. J. 46:1168–1172.

----, ----, and R.L. Blevins. 1986. Soil nitrogen availability after long-term continuous no-tillage and conventional corn production. Soil Sci. Soc. Am. J. 50:1206–1210.

Smith, M.S., and R.L. Blevins. 1987. Effect of conservation tillage on biological and chemical soil conditions. p. 149–166. *In* T.J. Logan et al. (ed.) Effects of conservation tillage on groundwater quality: Nitrates and pesticides. Lewis Publ., Chelsea, MI.

Smith, S.J., L.B. Young, and G.E. Miller. 1977. Evaluation of soil mineralization potentials under modified field conditions. Soil Sci. Soc. Am. J. 41:74–76.

Stanford, G. 1982. Assessment of soil nitrogen. *In* F.J. Stevenson (ed.) Nitrogen in agricultural soils. Agronomy 22:651–688.

Tisdall, J.M., and J.M. Oades. 1982. Organic matter and water-stable aggregates in soils. J. Soil Sci. 33:141–163.

Wilhelm, W.W., J.W. Doran, and J.F. Power. 1986. Corn and soybean yield response to crop residue management under no-tillage production systems. Agron. J. 78:184–189.

----, L.N. Mielke, and C.R. Fenster. 1982. Root development of winter wheat as related to tillage practice in western Nebraska. Agron. J. 74:85–88.

Winterlin, M.L., T.M. McCalla, and R.E. Luebs. 1958. Stubble mulch tillage versus plowing with nitrogen fertilization with regard to nutrient uptake by cereals. Agron. J. 50:241–243.

7 The Role of Nonleguminous Cover Crops in the Efficient Use of Water and Nitrogen[1]

M. G. Wagger

North Carolina State University
Raleigh, North Carolina

D. B. Mengel

Purdue University
West Lafayette, Indiana

The importance of nonleguminous cover crops in soil erosion control and enhancement of soil productivity via organic matter maintenance or even buildup has long been recognized in agriculture. Commonly used nonleguminous cover crops are the small grains and include rye (*Secale cereale* L.), wheat (*Triticum aestivum* L.), oat (*Avena sativa* L.), and barley (*Hordeum vulgare* L.). In some areas, perennial sod and annual forage crops have been used to provide vegetative cover (Box et al., 1980; Wilkinson et al., 1987). For the purpose of this discussion, soil protection provided by a living vegetative cover between periods of regular crop production will serve as a basis for defining cover crop systems and will, therefore, not include similar protection offered by various crop-residue management systems such as doublecropping or ecofallow.

With the refinement of conservation tillage technology, new cropping strategies have evolved with regard to cover crop management. Concomitant with these alternative management systems is the increasingly important need for evaluating current production schemes from the perspective of resource-use efficiency as a means to increase and sustain profitability for farmers and also to safeguard the environment. In this context, the components of a nonleguminous cover crop management system may offer a complex set of interrelated factors that affect plant growth and development and, in turn, final crop yield.

Two principal factors are water and N—the dynamic interactive effects of each that are often difficult to distinguish and evaluate. Moreover, since

[1] Contribution from the Dep. of Crop Science, North Carolina State Univ., Paper 11393 of the Journal Series of the North Carolina Agric. Res. Serv., Raleigh, NC 27695-7601, and the Agronomy Dep., Purdue Univ., West Lafayette, IN 47907.

Cropping Strategies for Efficient Use of Water and Nitrogen, Special Publication no. 51.

nonleguminous cover crop systems are adapted to a wide range of soil and climatic environments, the aforementioned interactive effects can vary considerably (particularly as related to soil water availability) between geographic regions. This chapter will attempt to outline current understanding and future research needs with regard to soil-plant-water relations and N dynamics in nonleguminous cover crop systems.

SOIL-PLANT-WATER RELATIONS

Winter annual grass cover crops are generally established in early fall by planting into prepared seedbeds, by interseeding into growing crops, or by no-till seeding into the previous crop's residue. With adequate growing conditions during the fall and early winter months, cover crops can often provide sufficient vegetative cover to protect the soil against wind and water erosion between principal crops. Soil moisture storage and subsequent availability are also affected by the presence of a vegetative cover. Potential benefits and limitations with regard to soil-moisture availability will be addressed in the following discussion.

Early Season Water Storage

Patterns of dry-matter accumulation and water use by cover crops change quite rapidly during early to mid-spring preceding spring-planted crops (Sprague et al., 1963; Utomo, 1986; Wagger, 1987). On humid-area soils characterized by moderate to high water-holding capacities, soil water depletion by actively growing cover crops prior to planting is generally not a critical factor. Although not well documented, field observations generally indicate that transpiration and soil drying provided by nonleguminous cover crops can often benefit more timely planting operations in this environment. By way of contrast, limited rainfall areas such as the Great Plains are not well-suited for cover cropping because of the need to store and conserve moisture (via precipitation) for the succeeding summer crop.

Cover crop management for soil moisture modification can be more crucial, however, on soils with low water-holding capacity or shallow rooting depth due to root restrictive layers. Campbell et al. (1984a), working with the Norfolk soil series (fine-loamy, siliceous, thermic Paleudults), a predominant Coastal Plain soil found in the southeastern USA, measured profile water content (0–60 cm) 15 d after corn (*Zea mays* L.) planting and found significantly greater soil-water depletion for a rye cover chemically or mechanically (disked) killed at planting compared to a conventional tillage system with no cover crop (Table 7–1). Paralleling these results were reductions in early season growth and grain yield. In another study with full-season soybean (*Glycine max* L. Merr.), water extracted by a rye cover crop at the time of desiccation was 2.5 cm (in the upper 37 cm soil) in the 25 d following disking for the conventional tillage system (Campbell et al., 1984b). However, even though dry soil conditions delayed germination and limited early season

Table 7-1. Effects of rye cover crop and tillage system on soil water and corn grain yield. Adapted from Campbell et al. (1984a).

Cover crop management	Soil water content by depth, cm				Grain yield
	0-15	15-30	30-45	45-60	
	%, w/w				Mg ha^{-1}
Conventional tillage (no cover)	8.9	9.9	19.4	21.4	6.16a*
Cover crop disked 1 d before planting	5.7	6.6	15.7	19.5	5.49ab
Cover crop killed (herbicide after planting)	5.4	5.6	13.7	18.5	4.93b
Cover crop strip killed (herbicide, 50% cover)	4.6	4.6	14.2	18.9	5.04b
Cover crop strip killed (mechanical, 50% cover)	2.4	3.9	14.1	18.2	3.92c

* Means followed by the same letter are not significantly different at $P \leq 0.05$.

growth of soybean planted into the rye cover, higher yields were ultimately produced in the cover crop system as a result of water conservation during a late-season drought.

The practice of in-row subsoiling to loosen highly compacted soil layers and cover management (removed vs. unremoved) were examined by Ewing et al. (1986) for North Carolina corn production. Mechanical harvesting 1 to 2 d prior to corn planting constituted the removed cover treatments, leaving approximately 5 cm of plant stubble. Soil water depletion by wheat and crimson clover (*Trifolium incarnatum* L.) in unremoved cover treatments significantly slowed emergence and subsequent vegetative corn development. While subsoiling proved to be the overriding factor determining final grain yield, yields were still reduced as a result of subsoil moisture drawdown prior to planting when the cover crop was unremoved, even with in-row subsoiling. On soils such as described above, dry weather conditions and erratic distribution of summer rainfall often fails to recharge profile soil moisture during the growing season, a condition exacerbated by the presence of live vegetative cover prior to planting the summer crop. Minimizing the effects of soil-water depletion by nonleguminous cover crops can be achieved by early herbicide desiccation, animal grazing, or mechanical harvesting (Campbell et al., 1984a).

Soil Moisture-Temperature Interaction

Numerous studies have delineated the influence of surface residue on soil temperature, whereby soils with residue cover maintain lower soil temperatures than soils with residues removed or incorporated by tillage (Moody et al., 1963; Cruz, 1982; Utomo et al., 1987). These temperature differences and their effect on crop growth and development can vary greatly with latitude and crop/planting date considerations.

Surface cover reduces the quantity of direct solar radiation reaching the soil surface (as well as reflecting radiation back to the atmosphere and providing a layer of insulation) and helps control evaporative losses of soil water, primarily before full-canopy development (Phillips, 1984a). Van Wijk et al. (1959) examined the consequences of mulched (oat straw) and unmulched soils on soil temperature in the root zone and early corn growth at four geographical locations in the eastern USA (Iowa, Minnesota, Ohio, and South Carolina). Single degree reductions (°C) resulted in small reductions in corn growth for mulched treatments at the southern location, while the same 1° reduction at the northern location markedly reduced corn growth. Soil temperatures at these locations were below optimum for corn-seedling growth and occasionally approached the critical temperature. The authors concluded that with a surface mulch present, the accompanying insulating effect and greater albedo was of more consequence in regulating soil temperature than the cooling effect of soil water evaporation.

Separating temperature and soil water effects under field conditions is difficult. Soil temperature and early season growth differences are not always associated with yield responses. It has often been observed that the early season growth depression accompanying surface residue environments reverses itself (greater growth and yields) after soil temperatures have sufficiently increased, presumably due to increased soil water availability under mulched conditions. Results from a growth chamber experiment demonstrated that corn germination rate was strongly influenced by soil water at a soil temperature of 31 °C, but virtually not at all of 15 °C (Cutforth et al., 1985). In another study, Cutforth et al. (1986) showed decreased root growth rate attributed to decreases in soil temperature and soil water. Sensitivity to soil water levels decreased with decreasing soil temperature and overall relationships were hybrid dependent.

Soil Moisture Conservation and Crop Water Use

Conservation of soil moisture by nonleguminous cover crops during the summer-growing season is due to reduced water evaporation from the soil surface prior to full-canopy establishment and also to decreased rainfall runoff and higher infiltration rates (Jones et al., 1969; Blevins et al., 1971; Phillips, 1984b). Experiments by Moschler et al. (1967) revealed higher surface soil moisture (0–15 cm), particularly during the first half of the growing season, under corn following a rye cover than under conventional tillage corn. Over 13 site-years, the rye cover/no-till corn system produced an average of 44% more grain yield than the conventional system in four comparisons and was comparable in the other nine. They concluded that the soil moisture conservation benefit derived from cover cropping was most likely to improve grain yield for droughts of short duration.

Cover management, such as removal for forage or using the cover crop for grazing, can alter the dynamics of soil water conservation and use. Soil water use by corn and soybean during periods of severe moisture stress indicated that rooting depth and activity were increased in a rye cover compared to rye stubble (Gallaher, 1977). Water conserved on this typical Piedmont soil (Cecil sl, clayey, kaolinitic, thermic Typic Hapludult) by the rye

Table 7-2. Tillage and winter crop management effects on corn and grain sorghum yields in the Southern Piedmont. Adapted from Nelson et al. (1977).

Principal crop management			Four-yr mean grain yield	
Tillage system	Planting date	Winter crop/management	Corn	Grain sorghum
			—— Mg ha^{-1} ——	
Conventional	30 April	Wheat/incorporated	7.5	4.1
No-till	30 April	Wheat/forage harvest	8.0	4.2
	30 April	Barley/forage harvest	8.3	4.6
	4 June	Wheat, barley/grain harvest	6.3	2.5

cover was calculated to withstand short-term droughts of approximately 7 d during the early growing season. This increase in soil moisture conserved, due to the presence of a mulch, was reflected in greater soil water use by root activity at deeper soil depths during critical grain-filling periods, as evidenced by grain yield increases of 46 and 30% for corn and soybean, respectively. Similar corn yield reductions due to rye cover removal just prior to planting were reported by Moschler et al. (1967) for several locations. Nelson et al. (1977), evaluating double-cropping alternatives for the Southern Piedmont, demonstrated the effectiveness of wheat and barley cover crops when harvested as forage for improving corn and grain sorghum [*Sorghum bicolor* (L.) Moench] yields in a no-till system over those in a conventional tillage system (Table 7-2). Harvesting the cover crops for grain led to a 5-week delay in the summer crop planting, along with substantial grain yield reductions.

Published reports dealing with nonleguminous cover crop effects on infiltration are extremely scarce, as most deal with management aspects of the previous crop residue or with somewhat artificially imposed systems designed to simulate effects of residue-loading rate. Nevertheless, general principles governing infiltration dynamics pertain for cover crop systems in humid environments. Surface residue enhances infiltration primarily by minimizing soil detachment (and thereby crusting and seal development) through the action of raindrop impact (Allmaras et al., 1985). Using simulated rainfall, Triplett et al. (1968) measured increased infiltration rates with increasing rate of corn residue cover on a Wooster silt loam (fine-loamy, mixed, mesic Typic Fragiudalf) with 1 to 2% slope. After 1 h, infiltration (cm) was plowed-bare, 1.80; no-tillage bare, 1.22; no-tillage with 40% surface residue cover, 2.34; and no-tillage with 80% surface residue cover, 4.39. Improved soil-structural stability was attributed to the positive influence of residue cover on infiltration. A threefold difference in mean infiltration rate for no-tillage vs. plowed treatments (0.91 vs. 0.31 cm min^{-1}) was reported by Lal (1976) for various cropping sequences on Alfisols in western Nigeria. Accompanying these findings were substantial reductions in runoff and soil loss in the mulched plots.

NITROGEN DYNAMICS

Nonleguminous cover crops can have a number of effects on the N relations in agricultural soils and, in turn, on subsequent crop N nutrition. These effects include:

1. A reduction in the supply of soil inorganic nitrogen due to cover crop uptake. By acting as an N sink, cover crops may or may not ultimately compete with the principal crop for available N.
2. The subsequent mineralization of cover crop N during the process of decomposition.
3. The net immobilization of fertilizer N during the decomposition of wide C/N ratio cover crops.
4. Avenues of N loss (leaching and/or denitrification) due to cover crop influence on soil water/N utilization relations. Conversely, cover crops may have a positive effect on summer crop N utilization under relatively droughty growing conditions.

Based on the above mechanisms, a nonleguminous cover crop could either increase the N availability to a subsequent crop by the trapping or capture of residual soil N, or could reduce N availability through competition or immobilization. These avenues of availability will be dependent on the cover crop grown and how it is managed. Nevertheless, while our intuitive understanding of these potential effects is generally agreed upon by scientists, good examples from published data are somewhat scarce. The following discussion will attempt to summarize the research available, and draw appropriate inferences where possible.

Nitrogen Uptake by Nonleguminous Cover Crops

Nitrogen accumulation has been studied in different environments and management systems in both wheat and rye cover crops. Where wheat or rye was planted after corn, and with no additional fertilizer N, cover crop uptake ranged from 12 to 91 kg of N ha^{-1} (Table 7-3). In most cases, uptake

Table 7-3. Amounts of residual/mineralized N taken up by unfertilized wheat or rye cover crops.

Experiment location	Reference	Site-years	Cover crops	Nitrogen content in vegetation	
				Range	$\bar{x}$
				—kg of N ha^{-1}—	
Alabama	Brown et al., 1985	2	Rye	19-36	24
Georgia	Hargrove, 1986	3	Rye	31-47	38
Indiana	Pelchat, 1986	5	Wheat	12-36	24
Kentucky	Ebelhar et al., 1984	2	Rye	--	36
Maryland	Decker et al., 1987	4	Wheat/barley	41-91	64
New York	Scott et al., 1987	3	Rye	16-26	21
North Carolina	Wagger, 1987	2	Rye	--	68
			$\bar{x}$	24-47	39

was in the range of 25 to 45 kg of N ha^{-1}. The exact amount of N taken up was a function of residual N remaining and the rate of N mineralization in that soil system.

Indirect examples of cover crops serving as a sink for residual N can be found in experiments conducted in Indiana and Kentucky. An Indiana study (Pelchat, 1986) demonstrated that a wheat cover crop in continuous no-till corn accumulated 12 to 36 kg of N ha^{-1}, depending on the site and year (Table 7-3). The amount of N applied to corn the previous year also had some effect on the N recovered by wheat (Table 7-4). Increasing N applications to the previous corn crop from 0 to 180 kg of N ha^{-1} resulted in a N uptake increase of 16 kg ha^{-1} by the subsequent cover crop. A similar increase in cover crop dry weight was also observed. This slight increase in N uptake by the cover crop reflects the rather modest amounts of residual N present in the fall where currently recommended N rates are used.

Research in Kentucky showed the N content of a rye cover crop increasing from 26 to 66 kg of N ha^{-1} as the N application to the previous corn crop increased from 0 to 170 kg of N ha^{-1} (Utomo, 1986). In another Kentucky study no measurements of cover crop N uptake were made, however, rye cover crop dry matter increased with each increment of fertilizer N to the previous corn crop (Saul, 1986). All three of these studies would suggest that nonleguminous cover crops can be utilized as a means of trapping residual N present in soils after the harvest of crops such as corn and sorghum. Thus, the potential exists for cover crops such as wheat and rye to be used as excellent tools in minimizing the effects of crop fertilization and residual N on the quality of ground and surface waters, and thereby preserving this N for use by subsequent grain crops. Hubbard et al. (1986) illustrated this concept by comparing model simulations of different N management systems on NO_3-N movement into shallow groundwater. The CREAMS model (Knisel, 1980), a field-scale model for chemicals, runoff, and erosion from agricultural management systems, was used to predict 20-yr NO_3-N loads from a Georgia Coastal Plain sand for two N-fertilizer rates (168 and 336 kg ha^{-1} yr^{-1}) and application methods (sidedressed and fertigation) with and without a winter cover of ryegrass (*Lolium multiflorum* Lam.). Total cumulative leaching increased with N rate and decreased with both winter

Table 7-4. Effect of prior fertilization on N uptake by subsequent cover crops.

Location	Reference	Cover crop	Previous N application	Cover crop yield	Nitrogen uptake
			kg ha^{-1}		
Indiana	Pelchat, 1986	Wheat	0	1600	18
			180	2970	34
Kentucky I	Utomo, 1986	Rye	0	2600	26
			170	6100	66
Kentucky II	Saul, 1986	Rye	0	1720	--
			75	2160	--
			150	2320	--
			225	2940	--

cover and fertigation. Averaged across N rates, the mean total NO_3-N load also decreased more with a winter cover than with fertigation (122 vs. 131 kg ha^{-1} yr^{-1}).

Unfortunately, cover crops can also compete with subsequent crops for fertilizer N. In an Indiana study, N fertilizer was applied in early April into a growing rye cover crop which was then allowed to grow about 4 weeks prior to corn planting. Nitrogen uptake by the rye cover increased from 71 to 158 kg of N ha^{-1} as fertilizer N rate increased from 0 to 201 kg of N ha^{-1} (S.A. Barber, 1987, personal communication). The yield of corn was reduced at both N rates where a cover crop was grown, and leaf N concentration indicated a reduction in N availability to the corn (Table 7–5). These results indicate the need for management systems that ensure adequate N availability to succeeding crops in nonleguminous cover crop systems.

Availability of Cover Crop Nitrogen

The effects of nonleguminous cover crops on the succeeding principal crop with regard to water relations was discussed previously; however, the effects can also be due to N dynamics in the cropping system. Numerous researchers have compared the N response of corn with and without cover crops (Ebelhar et al., 1984; Pelchat, 1986; Barber, 1987, unpublished; Decker et al., 1987; Tyler et al., 1987). In comparison to conventional tillage systems, a common trend in most of these studies has been a reduction in corn yield and N uptake and/or leaf N concentration where no N fertilizer was applied, and nonleguminous cover crops were grown (Table 7–6). However, in many but not all cases, yields were comparable with and without cover crops where recommended N rates were applied.

There are a number of possible explanations for these observations. The rate of N release from decomposing cover crop residues will be important in determining N available for subsequent crop growth. Factors affecting N turnover, such as the residue C/N ratio and the stage of growth termination will govern whether the cover crop acts as an N source or sink to the summer crop. Environmental conditions, primarily weather-related factors such as temperature or precipitation, can also influence the rate of residue decomposition.

Table 7–5. Fertilizer N uptake by a rye cover crop and subsequent corn earleaf N concentration and grain yield (Barber, 1987, unpublished data).

Cover crop	Nitrogen rate	Nitrogen uptake	Corn earleaf N	Corn grain yield
	kg of N ha^{-1}		g kg^{-1}	Mg ha^{-1}
None	0	--	17.1	6.2
	201	--	26.4	8.9
Rye	0	71	14.5	4.0
	201	158	23.4	7.9

Wagger (1987) examined the consequences of cover crop desiccation relative to corn planting (early kill/early plant, early kill/late plant, and late kill/late plant) on subsequent residue N turnover and corn uptake of N. A 2-week delay in desiccation of a rye cover crop resulted in cover crop N content increasing approximately 13% (63 vs. 71 kg of N ha^{-1}), while the C/N ratio increased from 32:1 to 38:1. Differences in N-release patterns were also readily apparent, as the early kill date depicted a faster rate of N turnover over the entire growing season, with differences more pronounced in a season marked by relatively dry conditions. Nevertheless, all desiccation/corn planting combinations in the rye cover crop system were characterized by a net N immobilization. These results are not surprising when one considers that a C/N ratio of 25 or greater will cause some of the inorganic nitrogen to be immobilized (Allison, 1966). The C/N ratio of plant residues has frequently been used as an indicator for predicting the rate of decomposition, yet is not the sole determinant. In the above-cited study, increases in the cellulose and lignin fractions were also associated with the late desiccation.

Work by Utomo (1986) in Kentucky shows the importance of supplemental N on the rate of residue decomposition. Rye decomposition was measured over a 14-week period where corn had received 0 or 170 kg of N ha^{-1}, both in the previous and test years. Large differences in residue biomass were initially observed. Where N was applied in the previous year, a twofold increase in cover crop residue production existed compared to plots where no N was applied. The N concentration of the residue was similar in both situations, however. Even though initial C/N ratios were similar (ca. 27:1), the rate of residue disappearance was much greater in the fertilized plots. Addition of fertilizer N enhanced the rye decomposition by providing more available N for microbial decomposition as well as decreasing the C/N ratio.

Table 7-6. Cover crop effects on subsequent corn N utilization and growth.

Location	Reference	Cover crop/ Tillage	Nitrogen applied to corn	Corn earleaf N	Corn N uptake	Grain yield
			kg ha^{-1}	g kg^{-1}	— kg ha^{-1}	—
Indiana	Pelchat, 1986	None/plowed	0	14.2	--	3760
		None/no-till	0	8.9	--	1750
		Wheat/no-till	0	9.6	--	2220
		None/plowed	180	26.8	--	7330
		None/no-till	180	25.0	--	7430
		Wheat/no-till	180	24.5	--	6780
Kentucky	Ebelhar et al., 1984	None/no-till	0	14.9	--	3800
		Rye/no-till	0	14.3	--	4000
		None/no-till	100	22.5	--	6800
		Rye/no-till	100	21.5	--	7600
Maryland	Decker et al., 1987	None/no-till	0	--	99	5050
		Wheat, barley/no-till	0	--	65	3220
		None/no-till	90	--	158	8040
		Wheat, barley/no-till	90	--	131	7240

Evidence of residual and fertilizer N immobilization during microbial decomposition of rye (Hargrove, 1986) and wheat (Tyler et al., 1987) cover crop residues was typified in N rate experiments conducted in Georgia (grain sorghum) and Tennessee (corn), respectively. In each case (sorghum grain N content and corn grain yield as functions of fertilizer N rate), a lower y-intercept was calculated for regression equations for rye and wheat cover crops compared to a fallow (no cover) system. It should be noted, however, that environmental conditions can present problems in separating N dynamics from soil moisture effects. Wagger (1986) observed greater fertilizer N-utilization efficiency by corn planted into a rye cover at 90 and 180 kg of N ha^{-1} rates than the corresponding fallow treatments, while at 0 N applied, immobilization was manifested in the rye cover treatment. These responses were likely due to a significant mulch effect (soil water) under the extremely dry conditions that prevailed during the study.

Fertilizer N placement is important where large amounts of wide C/N ratio residue are present. Barber (1987, personal communication) found a reduction in both corn yields and leaf N concentrations when fertilizer N was applied early enough to allow significant N utilization by the rye cover crops. Similar situations would exist if the fertilizer was applied directly to, or mixed with the residue to enhance immobilization. Knifing the fertilizer below the residue would reduce immobilization and enhance the availability of fertilizer N to the target crop, as commonly seen in high residue no-till systems (Mengel et al., 1982; Rice and Smith, 1984).

Nitrogen and Cover Crop Management to Enhance Nitrogen-use Efficiency

Published information regarding N-use efficiency relationships in nonleguminous cover crop systems is limited, as most work has focused on these concerns in previous-crop residue systems. General concepts are still relevant, however, for most of the N processes/mechanisms with wide applicability to biological systems. It follows that a number of management factors should be considered when designing a cropping system that includes cover crops to ensure adequate N-use efficiency and optimum crop growth. These would include the type of cover crop (leguminous vs. nonleguminous), stage at which the growth of the cover crop is terminated, tillage system, and fertilizer N required—with consideration for appropriate N source, timing and method of application. Other inputs such as nitrification inhibitors should also be considered, particularly if the cover crop would have a significant effect on water relations which, in turn, could influence N losses through leaching or denitrification.

As previously discussed, the addition of large amounts of wide C/N ratio residues can create conditions conducive to N immobilization during the decomposition of those residues. Thus, managing the cover crop to reduce this potential could prove beneficial. Chemically killing the cover crop at a relatively early growth stage, with a resultant lower C/N ratio, could help reduce the N immobilization potential.

Physical placement of fertilizer in relation to the residue could also prove important. Fertilizer N applied directly to residue covering the soil surface could enhance both ammonia volatilization (Keller and Mengel, 1986) and N immobilization (Rice and Smith, 1984). Separating the fertilizer from the residue, or minimizing the amount of fertilizer/residue contact, could also enhance the efficiency of N use.

FUTURE RESEARCH NEEDS

Nonleguminous cover crops offer the opportunity to enhance N-use efficiency, conserve soil and water, and preserve water quality by acting as sink for residual N that might otherwise leave the soil system. This would also preserve an N resource for use by subsequent crops. However, if we are to effectively utilize this tool for environmental enhancement and improved cropping efficiency, our present knowledge regarding cover crop effects on water and N dynamics in soils needs to be expanded. Under this framework, we must learn to effectively manage nonleguminous cover crops over a wider range of conditions. A number of particular research needs exist, including, but not limited to, the following:

- Definition of the effects of cover crops on run-off, infiltration, and water storage.
- Clarification of the role of cover crops as sinks for residual N (to include long-term effects on soil organic carbon and nitrogen levels), and a better understanding of the management of cover crops in this role to ensure that the residual N trapped is readily available to subsequent summer crops.
- Better understanding of N-fertilizer management in cover crop systems to maximize N-use efficiency.

REFERENCES

Allison, F.E. 1966. The fate of nitrogen applied to soils. Adv. Agron. 18:219–258.

Allmaras, R.R., P.W. Unger, and D.W. Wilkins. 1985. Conservation tillage systems and soil productivity. p. 358–404. *In* R.F. Follett and B.A. Stewart (ed.) Soil erosion and crop productivity. ASA, CSSA, and SSSA, Madison, WI.

Blevins, R.L., D. Cook, S.H. Phillips, and R.E. Phillips. 1971. Influence of no-tillage on soil moisture. Agron. J. 63:593–596.

Box, J.E., Jr., S.R. Wilkinson, R.N. Dawson, and J. Lozachyn. 1980. Soil water effects on no-till corn production in strip and completely killed mulches. Agron. J. 72:797–802.

Brown, S.M., T. Whitwell, J.T. Touchton, and C.H. Burmester. 1985. Conservation tillage systems for cotton production. Soil Sci. Soc. Am. J. 49:1256–1260.

Campbell, R.B., D.L. Karlen, and R.E. Sojka. 1984a. Conservation tillage for maize production in the U.S. Southeastern Coastal Plain. Soil Tillage Res. 4:511–529.

----, R.E. Sojka, and D.L. Karlen. 1984b. Conservation tillage for soybean in the U.S. Southeastern Coastal Plain. Soil Tillage Res. 4:531–541.

Cruz, J.C. 1982. Effect of crop rotation and tillage systems on some soil physical properties, root distribution, and crop production. Ph.D. diss., Purdue Univ., West Lafayette, IN (Diss. Abstr. 83-00897).

Cutforth, H.W., C.F. Shaykewick, and C.M. Cho. 1985. Soil water-soil temperature interactions in the germination and emergence of corn (*Zea mays* L.). Can. J. Soil Sci. 65:445–455.

----, C.F. Shaykewick, and C.M. Cho. 1986. Effect of soil water and temperature on corn (*Zea mays* L.) root growth during emergence. Can. J. Soil Sci. 66:51–58.

Decker, A.M., J.F. Holderbaum, R.F. Mulford, J.J. Meisinger, and L.R. Vough. 1987. Fall-seeded legume nitrogen contributions to no-till corn production. *In* J.F. Powers (ed.) The role of legumes in conservation tillage systems: The proceedings of a national conference. Soil Conserv. Soc. Am., Ankeny, IA.

Ebelhar, S.A., W.W. Frye, and R.L. Blevins. 1984. Nitrogen from legume cover crops for no-tillage corn. Agron. J. 76:51–55.

Ewing, R.P., M.G. Wagger, and H.P. Denton. 1986. Effects of subsoiling and cover crop on growth and yield of corn. p. 111. *In* Agronomy abstract. ASA, Madison, WI.

Gallaher, R.N. 1977. Soil moisture conservation and yield of crops no-till planted in rye. Soil Sci. Soc. Am. J. 41:145–147.

Hargrove, W.L. 1986. Winter legumes as a nitrogen source for no-till grain sorghum. Agron. J. 78:70–74.

Hubbard, R.K., G.J. Gascho, J.E. Hook, and W.G. Knisel. 1986. Nitrate movement into shallow ground water through a Coastal Plain sand. Trans. ASAE 29:1564–1571.

Jones, J.N., Jr., J.E. Moody, and J.H. Lillard. 1969. Effects of tillage, no tillage, and mulch on soil water and plant growth. Agron. J. 61:719–721.

Keller, G.D., and D.B. Mengel. 1986. Direct measurements of ammonia volatilization loss from surface applied nitrogen in no-till corn. Soil Sci. Soc. Am. J. 79:1026–1029.

Knisel, W.G. (ed.) 1980. CREAMS: A field-scale model for chemicals, runoff, and erosion from agricultural management systems. Conserv. Res. Rep. 26. USDA, Washington, DC.

Lal, R. 1976. No-tillage effects on soil properties under different crops in western Nigeria. Soil Sci. Soc. Am. J. 40:762–768.

Mengel, D.B., D.W. Nelson, and D.M. Huber. 1982. Placement of nitrogen fertilizers for no-till and conventional till corn. Agron. J. 74:515–518.

Moody, J.E., J.N. Jones, Jr., and J.H. Lillard. 1963. Influence of straw mulch on soil moisture, soil temperature and the growth of corn. Soil Sci. Soc. Am. Proc. 27:700–703.

Moschler, W.W., G.M. Shear, D.L. Hallock, R.D. Sears, and G.D. Jones. 1967. Winter cover crops for sod-planted corn: Their selection and management. Agron. J. 59:547–551.

Nelson, L.R., R.N. Gallaher, R.R. Bruce, and M.R. Holmes. 1977. Production of corn and sorghum grain in double-cropping systems. Agron. J. 69:41–45.

Pelchat, J.A. 1986. Effects of tillage and winter cover crops on nitrogen requirements of corn in Indiana. Ph.D. diss. Purdue Univ., West Lafayette, IN (Diss. Abstr. 87-00940).

Phillips, R.E. 1984a. Effects of climate on performance of no-tillage. p. 11–41. *In* R.E. Phillips and S.H. Phillips (ed.) No-tillage agriculture principles and practices. Van Nostrand Reinhold Co., New York.

----. 1984b. Soil moisture. p. 66–86. *In* R.E. Phillips and S.H. Phillips (ed.) No-tillage agriculture principles and practices. Van Nostrand Reinhold Co., New York.

Rice, C.W., and M.S. Smith. 1984. Short-term immobilization of fertilizer nitrogen at the surface of no-till and plowed soils. Soil Sci. Soc. Am. J. 48:295–297.

Saul, M.R. 1986. Effect of tillage, nitrogen fertilization, and cover crop management in continuous corn production. M.S. thesis. Univ. of Kentucky, Lexington.

Scott, T.W., J. Mt. Pleasant, R.F. Burt, and D.J. Otis. 1987. Contributions of ground cover, dry matter, and nitrogen from intercrops and cover crops in a corn polyculture system. Agron. J. 79:792–798.

Sprague, M.A., M.M. Hoover, Jr., M.J. Wright, H.A. MacDonald, B.A. Brown, A.M. Decker, J.B. Washko, K.E. Varney, and V.G. Sprague. 1963. Seeding management of grass-legume associations in the Northeast. New Jersey Agric. Exp. Stn. Bull. 804.

Triplett, G.B., Jr., D.M. Van Doren, and B.L. Schmidt. 1968. Effect of corn (*Zea mays* L.) stover mulch on no-tillage corn yield and water infiltration. Agron. J. 60:236–239.

Tyler, D.D., B.N. Duck, J.G. Graveel, and J.F. Bowen. 1987. Estimating response curves of legume nitrogen contribution to no-till corn. *In* J.F. Powers (ed.) The role of legumes in conservation tillage systems: The proceedings of a national conference. Soil Conserv. Soc. Am., Ankeny, IA.

Utomo, M. 1986. Role of legume cover crops in no-tillage and conventional tillage corn production. Ph.D. diss., Univ. of Kentucky, Lexington.

----, R.L. Blevins, and W.W. Frye. 1987. Effect of legume cover crops and tillage on soil water, temperature, and organic matter. p. 5–6. *In* J.F. Power (ed.) The role of legumes in conservation tillage systems: Proceedings of a national conference. Soil Conserv. Soc. Am., Ankeny, IA.

Van Wijk, W.R., W.E. Larson, and W.C. Burrows. 1959. Soil temperature and the early growth of corn from mulched and unmulched soil. Soil Sci. Soc. Am. Proc. 23:428–434.

Wagger, M.G. 1986. Cover crop management and N rate effects on corn growth and yield. p. 124. *In* Agronomy abstract. ASA, Madison, WI.

----. 1987. Timing effects of cover crop dessication on decomposition rates and subsequent nitrogen uptake by corn. p. 35–37. *In* J.F. Power (ed.) The role of legumes in conservation tillage systems: The proceedings of a national conference. Soil Conserv. Soc. Am., Ankeny, IA.

Wilkinson, S.R., O.J. Devine, D.P. Belesky, J.W. Dobson, Jr., and R.N. Dawson. 1987. No-tillage intercropped corn production in tall fescue sod as affected by sod-control and nitrogen fertilization. Agron. J. 79:685–690.

8 Role of Annual Legume Cover Crops in Efficient Use of Water and Nitrogen

W. W. Frye, R. L. Blevins, M. S. Smith, and S. J. Corak

University of Kentucky
Lexington, Kentucky

J. J. Varco

Mississippi State University
Mississippi State, Mississippi

Crops grown in rainfed production systems are usually subjected to periods of drought stress of varying length and frequency during the growing season. The mulch formed by a cover crop with conservation tillage can eliminate some of these short periods of drought stress and lessen, but not eliminate, the effects of longer periods. Soil water conservation is one of the major benefits of conservation tillage, but this and other advantages of the practice are heavily dependent upon the presence of a vegetative mulch cover on the soil surface. The most effective cropping strategy to assure a mulch cover during the primary cropping season is to grow a cover crop during the dormant season or off-season, usually the winter. Winter cover crops of either grasses or legumes are compatible with conservation tillage methods and offer several important benefits that protect and improve the soil and can increase production efficiency. Until recently, cover crops used with conservation tillage have been mostly small grains, particularly wheat (*Triticum aestivum* L.) and rye (*Secale cereale* L.).

Since 1975, there has been renewed interest by both farmers and agronomists in cropping strategies that combine legume cover crops with conservation tillage. Kindled by the energy crisis and rising cost of N fertilizer in the 1970s, this interest has been fueled by the rising awareness of the need to control soil erosion, restore productivity of eroded soils, and improve efficiency in crop production.

A legume cover crop provides a substantial amount of biologically fixed N to the primary crop, as well as providing the other advantages offered by a nonlegume cover crop. In some cases, the amount of N provided by

[1] Contribution of the Dep. of Agronomy, Univ. of Kentucky, Lexington, KY 40546-0091.

Cropping Strategies for Efficient Use of Water and Nitrogen, Special Publication no. 51.

a legume cover crop may be adequate to produce optimum yields of nonleguminous crops, but in other cases, especially with corn (*Zea mays* L.), the legume N has to be supplemented with N fertilizer. However, N-fertilizer rates can be lowered appreciably while maintaining optimum economic yields.

A cover crop is not without liabilities. Chief among these are the cost of seeding and depletion of stored soil water. Seeding cost must be gauged against possible economic benefits such as increased yields or decreased N-fertilizer needs. Depletion of soil water tends to contradict the water conservation advantage mentioned above; however, the cover crop depletes soil water while it is still growing but conserves soil water after it is killed.

The objectives of this chapter are (i) to describe the methods of using annual legume winter cover crops in various cropping systems, mainly continuous grain; (ii) to evaluate the effects of these cover crops on soil water conservation and use; (iii) to examine the transformations and plant availability of N from several adapted species of annual legume cover crops; and (iv) to discuss approaches to determining the N-fertilizer requirement for a summer crop following a legume winter cover crop. The chapter emphasizes, although not restricted to, continuous grain production systems with no-tillage. Where possible, comparisons to conventional tillage (usually moldboard plowing and disking) will be made.

MANAGEMENT OF ANNUAL LEGUME WINTER COVER CROPS

Selecting the Cover Crop

Selecting a legume that is well adapted to the climate and management conditions is the first and most important decision. Hairy vetch (*Vicia villosa* Roth) and crimson clover (*Trifolium incarnatum* L.) appear to be the two best-adapted legume winter cover crops for the Southeast. They are easily established, reasonably winter hardy, break dormancy early in the spring, grow rapidly, assimilate high amounts of N, and result in good yield responses by summer crops. Hairy vetch is more winter hardy, but crimson clover accumulates its dry matter and N earlier in the spring.

Hairy vetch, being more winter hardy than crimson clover, appears to be particularly well adapted to the mid-section of the USA. In Maryland, Delaware, Kentucky, and Nebraska performance of hairy vetch has been consistently high in several experiments. It is probably the most widely adapted legume winter cover crop, being one of the most important cover crops throughout the Southeast (Smith et al., 1987) and performing fairly satisfactorily as far north as central New York (Scott and Burt, 1985). Touchton et al. (1982) and Hargrove (1986) in Georgia have generally obtained superior results with crimson clover. In contrast, crimson clover has not produced high N levels nor satisfactory grain-yield responses in Kentucky (Ebelhar et al., 1984). Bigflower vetch (*Vicia grandiflora* var. *Kitaibeliana* W. Koch) has been equal to hairy vetch in all respects in studies at Princeton, KY (Herbek et al., 1987), but its performance at Lexington, KY has been inferior to hairy vetch (Ebelhar et al., 1984; Utomo, 1986). Decker et al. (1987), in Maryland, obtained an excellent corn yield response following cover crops of Austrian winter pea [*Pisum sativum* sap. (L.) Poir] and with hairy vetch. Under dryland conditions of northern Idaho in 1979 and 1980, five lines of fall-planted (early September) Austrian winter pea produced an average of 10.8 Mg ha^{-1} of harvested dry matter containing 279 kg of N ha^{-1} when harvested in mid-June (Auld et al., 1982). Winter survival of Austrian winter pea has been so poor in Kentucky that we have concluded that it is unsuitable as a winter cover crop under our climate and management conditions. Subterranean clover (*Trifolium subterraneum* L.) has recently shown excellent potential as a winter cover crop in North Carolina and Georgia and common vetch (*Vicia sativa* L.) equalled hairy vetch in dry matter yield, but tended to accumulate slightly less N in Georgia (Hoyt and Hargrove, 1986). Wright and Coxworth (1987) obtained favorable results from the use of faba bean (*Vicia faba* L.) as a pulse crop with barley in northern Canada.

Hoyt and Hargrove (1986) listed adaptational characteristics of a number of other commonly planted legume species. Several of them have shown promise as cover crops, but have not yet been thoroughly tested.

Planting the Cover Crop

Planting time and method must be selected to match the weather conditions and cropping system. Cost of seeding is the main expense of a legume cover crop, and obtaining an adequate stand is essential to the effectiveness of the cover crop as a source of N and as a mulch.

There are several practical ways to establish annual legume cover crops into the cropping systems. Aerial interseeding without incorporation into a standing summer crop before harvest and just before the beginning of leaf-drop has been used most frequently in Kentucky. With corn, soybean [*Glycine max* (L.) Merr.], and grain sorghum [*Sorghum bicolor* (L.) Moench], this normally occurs early to mid-September. There are two keys to success in this method of planting—adequate soil moisture for germination and seedling growth and adequate plant residue to cover most of the seeds. The residue can be leaf-drop from the current crop or residue from previous crops. For these reasons, it is usually easier to establish cover crops in no-tillage fields than in conventional tillage fields using the aerial interseeding method.

Legume cover crops may also be planted with a no-tillage drill after the grain crop has been harvested. This method places the seeds into the soil below the crop residue. Thus, it has the advantage of placing the seeds into a more favorable environment for germination. A major disadvantage of this method is that planting must be delayed until after harvest. Later planting may make the cover crop more susceptible to winter-kill and decrease its yield and effectiveness.

We compared no-tillage drilling after corn harvest with aerial interseeding on the soil surface about 3 weeks before harvest (Table 8–1). Both methods were successful, but the later was superior regarding dry matter and N production. Brown et al. (1985) found a rather spectacular increase in both dry

Table 8-1. Effect of planting method on dry matter yield and N production of cover crops.

Cover crop	Dry matter yield		Nitrogen content		Reference
	IS†	D	IS	D	
	—Mg ha^{-1}—		—kg ha^{-1}—		
Hairy vetch	3.76	3.06	135	104	Frye et al. (1984),
Bigflower vetch	2.74	2.37	98	85	Princeton, KY;
Rye	2.31	2.14	21	19	avg. 4 yr
Hairy vetch	2.96	1.83	133	75	Brown et al. (1985),
Crimson clover	4.38	1.33	133	44	Alabama
Rye	2.84	1.90	27	21	

† IS = Interseeded early September before harvest in Kentucky and approximately 25 September before cotton defoliation in Alabama; D = Drilled after crops harvested—mid-October in Kentucky and early November in Alabama.

matter production and N accumulation where cover crops were interseeded compared with drilled following cotton (*Gossypium hirsutum* L.) (Table 8-1). Conversely, Decker et al. (1987) stated that seeding with a no-tillage drill after corn harvest was more successful than surface seeding. Our observations in Kentucky suggest that planting date may be a more important consideration than placement of seed.

Several other seeding methods have been reported to be successful. In Canada, Nanni and Baldwin (1987) seeded a legume cover crop behind the last cultivation of a corn crop using rolling packers attached behind the cultivator to obtain good seed-soil contact. Ngalla and Eckert (1987) drilled red clover (*Trifolium pratense* L.) into winter wheat in April or after wheat harvest in July. Corn was planted into the wheat-red clover residue late in April the following year. They reported that drilling into standing wheat in April decreased wheat yields (presumably due to damage from the drill) by about 4%, and they pointed out that the red clover could be established by broadcast seeding, making drilling unnecessary.

Once established, certain legumes are effective in self-reseeding systems. The self-reseeding technique of perpetuating a winter cover crop involves waiting to kill at least a portion of the cover crop until it has produced viable seed. This can be done by waiting until after seed production to plant the summer crop, or by planting the summer crop and leaving a strip of living cover crop between the rows until reseeding is complete. This strip can be killed after seed production using a shielded sprayer.

The major advantage of this practice is to avoid having to plant the legume cover crop each fall, thus saving the cost of annual seeding. Crimson clover was shown to be particularly well-adapted to self-reseeding systems by Touchton et al. (1982) in Georgia and Oyer and Touchton (1987) in Alabama. In Kentucky, bigflower vetch has shown excellent potential as a reseeding winter legume cover crop for no-tillage corn (Jauregui, 1986; Frye et al., 1983).

Table 8-2. Effect of kill date on dry matter yield and N content of winter cover crops. Adapted from Wagger (1987).

Cover crop	Dry matter†		Nitrogen content†	
	Early kill‡	Late kill	Early kill	Late kill
	Mg ha^{-1}		kg ha^{-1}	
Hairy vetch	3.36	5.38	134	192
Crimson clover	3.81	5.38	113	139
Rye	5.82	8.51	80	90

† Aboveground vegetation; avg. 1984 and 1985.
‡ Early kill—3rd week of April; Late kill—1st week of May.

Killing the Cover Crop

A legume winter cover crop may be used with any tillage system ranging from conventional tillage to no-tillage. If plowing or heavy disking is used to bury the cover crop in the spring, kill-down herbicides are not needed. If the cover crop is used to provide a surface mulch in a conservation tillage system, it is necessary to kill it with herbicides to prevent it from competing with the summer crop.

A number of herbicides have been proven satisfactory. Worsham and White (1987) and Dabney and Griffin (1987) listed several herbicides and mixtures of herbicides used in experiments with legume cover crops and show their comparative effectiveness.

Herbicides may be applied preplant or when the summer crop is planted. The main advantage of postponing spraying as late as possible is the additional dry matter and N production obtained from the legume. Better timing of N release to coincide with N needs of the summer crop could be another advantage of spraying at planting. Whether this is an advantage or not depends on when the crop's N need occurs and whether some tillage is to be used. As will be discussed later, tillage speeds up residue decomposition and N mineralization (Varco, 1986). Conversely, the main advantage of killing the cover crop earlier is the stored soil water that is saved. In addition, the N concentration of the legume may be greater when killed preplant than when killed at a more mature growth stage.

Wagger (1987), in North Carolina, found that considerable dry matter yield and N content were sacrificed when cover crops were killed in the 3rd week of April compared to the 1st week of May (Table 8-2). The effect was more pronounced for hairy vetch than for crimson clover. Crimson clover reaches peak production a few days ahead of hairy vetch. During about 2 weeks, hairy vetch produced about 2.00 Mg ha^{-1} dry matter and 58 kg of N ha^{-1}. Corresponding values for crimson clover were 1.57 Mg ha^{-1} dry matter and 26 kg of N ha^{-1}. This is consistent with our observations in Kentucky that winter legume cover crops make most of their growth during about 3 weeks prior to entering their reproductive stage.

Harvesting vs. Leaving the Cover Crop

Table 8-3 shows that harvesting the cover crop greatly decreased subsequent corn yields where no N fertilizer was applied. Where N fertilizer was applied, yields were usually decreased by harvesting the cover crop, but by a lesser amount than without fertilizer.

Leaving the cover crop as a mulch or green manure clearly is agronomically advantageous. Whether it would be economically advantageous depends on several economic factors, including the cost of N fertilizer, the value of the harvested cover crop, and the value of the additional corn yield.

SOIL WATER RELATIONSHIPS

Smith et al. (1987) indicated four ways in which legume winter cover crops may affect the plant-soil water relationships of the summer crop. They listed: (i) decreased evaporation due to the mulch formed by the cover crop, (ii) increased infiltration of rainfall, (iii) loss of stored soil water by transpiration of the cover crop, and (iv) altered water use by the summer crop. A vegetative mulch associated with conservation tillage is probably the most effective way of increasing the plant-available soil water in large-scale agricultural production. Mulches from annual legume winter cover crops improve soil water conditions by (i) decreasing surface runoff, (ii) increasing soil organic matter and improving soil structure, and (iii) decreasing evaporation losses.

Surface Runoff

Reducing surface runoff and increasing infiltration of rainfall is probably the most important way in which the cropping system, excluding the tillage system, affects water conservation (Frye et al., 1985a). That a cover crop decreases surface runoff of water has been shown repeatedly. Where

Table 8-3. Effect of harvesting legume cover crops on yield of corn grain.

		Cover crop residue mgt.				
		Removed		Left		
Cover crop	Tillage†	0‡	118	0	118	Reference
		— Grain yield, Mg ha^{-1} —				
Crimson clover	CT	4.1	8.2	5.1	7.8	Hargrove, 1982
	NT	5.7	6.9	8.1	8.3	
Hairy vetch	CT	5.8	7.9	7.9	8.3	
	NT	4.5	6.7	6.3	7.9	
		0†	90	0	90	
Crimson clover	NT	5.4	8.1	6.8	8.6	Holderbaum et al., 1987
Fallow	NT	2.9	6.5			

† CT = conventional tillage; NT = no-tillage. ‡ Fertilizer-N rate, kg ha^{-1}.

runoff is slowed by mulch cover, infiltration of water increases simply becuase it has more time to enter the soil. In addition to slowing the runoff, the mulch protects the soil from the impact of raindrops and reduces the breakdown of soil aggregates, thereby preventing the filling of macropores and soil crusting.

Table 8–4 shows the rainfall runoff losses measured on two soils in Kentucky. It is likely that the lower runoff with no-tillage compared to conventional tillage is a result of the mulch cover provided by the residue of the previous crop. Others have found similar relationships (Hale et al., 1984). Interestingly, a wheat cover crop with no-tillage did not decrease runoff loss over no-tillage without a cover crop at Princeton (Table 8–4). This suggests that residue from a cover crop may not improve runoff control with no-tillage over that obtained with residue from a previous summer crop.

Felton et al. (1987) stated that 90% cover was needed to achieve maximum infiltration into a dry, black earth soil in Australia. With 4 Mg ha^{-1} wheat stubble, infiltration was 80% of the rainfall. They suggested a target of 4 Mg ha^{-1} of mulch to minimize runoff.

Effects of Soil Organic Matter

Except in sandy soils, perhaps the most important effect of organic matter on soil water is through its influence on soil structure. Soil organic matter is generally considered to be the most effective stabilizer of soil aggregates (Elliott et al., 1987). Soils with stable, granular structure are more resistant to the destructive effects of raindrops than soils with weak structure; therefore, more soil pores remain open, and infiltration is faster. With more infiltration and less runoff from each rainfall event, the amount of plant-available water is greater and soil erosion is less.

Increasing tillage usually decreases the organic matter content of soils and decreasing tillage usually increases it. When a soil previously in bluegrass (*Poa pratensis* L.) pasture for about 50 yr was cropped for 10 yr in continuous corn with rye cover crops, the organic matter content decreased in the 0-

Table 8–4. Runoff losses as affected by tillage. Unpublished data from R. Blevins and W. Frye (Lexington, KY) and D. Ditsch and M. Rasnake (Princeton, KY).

	Runoff loss			
		Princeton		
Tillage	Lexington†	1985‡	1986‡	1987‡
	cm ha^{-1}			
Conventional tillage	1.6	7.7	--	--
Chisel-plow tillage	0.5	--	3.3	8.6
No-tillage, without cover crop	0.6	2.9	2.1	2.9
No-tillage, wheat cover crop	--	3.1	1.8	2.2

† Three-year average (1985–1987) from corn plots with tillage and planting perpendicular to 9% slope.
‡ 15 May through 28 Oct. 1985; October to October during 1986 and 1987. Soybean plots with tillage and rows up-and-down 7 to 9% slope; tillage done in fall.

to 5-cm depth of both no-tillage and conventional tillage soil relative to soil left under bluegrass sod (Blevins et al., 1983). The decrease was 13.1 g of C kg^{-1} in the no-tillage soil compared to 24.6 g of C kg^{-1} for the plowed soil.

Conservation tillage usually increases the organic matter content of soils that have been conventionally tilled for a long time (Elliott et al., 1987). Generally, the more the organic matter has been depleted, the more rapid the increase. No-tillage corn in a vetch and rye mulch increased the soil organic matter content of the Ap horizon from 15 to 26 g kg^{-1} during a 10-yr period in South Carolina (Beale et al., 1955). The organic matter content of the plowed check plots decreased from the initial 15 to 12 g kg^{-1} during the period. Their study also showed that the degree of soil structural aggregation and stability improved under the mulched treatments, but was reduced under the plowed check treatment.

Increasing the soil organic matter content has long been considered to be one of the important benefits of legume cover crops. Either legume cover crops or conservation tillage alone will increase the soil organic matter content over conventional tillage without cover crops, but the best results are usually obtained when the two practices are combined. Hargrove (1982) reported results from Georgia showing that organic carbon in the 0- to 7.5-cm depth of soil was 14.2 g kg^{-1} after 3 yr of conventionally tilled wheat/soybean and 24.2 g kg^{-1} under a no-tillage crimson clover/grain sorghum system. In this case, the results were thought to be due more to the greater amount of residue added to the soil by the crimson clover/grain sorghum system than to differences in tillage (Hargrove and Frye, 1987). This conclusion was supported by the fact that no-tillage wheat/soybean without clover was only 2 g kg^{-1} higher in organic carbon (16.1 g kg^{-1}) than conventionally tilled wheat/soybean.

By contrast, Utomo (1986) found a sizable increase in the organic carbon content with no-tillage compared to conventional tillage corn in Kentucky (Table 8-5). Differences among winter cover treatments of corn residue alone (fallow) and corn residue plus either hairy vetch or rye were small, indicating that the much greater amounts of residue from the hairy vetch and rye treatments did not accumulate as soil organic matter. The initial organic matter content was relatively high at about 20 g kg^{-1} organic carbon in the 0- to 7.5-cm depth of this soil. In such soils, the greatly altered soil

Table 8-5. Organic carbon in 0- to 7.5-cm depth of soil on 30 Sept. 1985 as affected by winter cover, N-fertilizer rate, and tillage treatment. Adapted from Utomo (1986).

	Fertilizer N, kg ha^{-1}			
	0		170	
Winter cover	CT†	NT	CT	NT
	organic carbon, g kg^{-1}			
Fallow	15.7d*	19.3c	16.8d	23.7d
Hairy vetch	16.0d	22.0b	16.9d	25.2a
Rye	16.3d	19.8c	15.2d	22.4b

* Values followed by different letters are significantly different ($P < 0.05$).
† CT = plowed and disked; NT = no-tillage.

environment produced by no-tillage will increase the soil organic matter content, whereas an increase in residue beyond that required to maintain the equilibrium level established by the new environment might not increase the organic matter level over the long term. This additional residue may be quite beneficial, however, in increasing infiltration, decreasing runoff, improving soil erosion control, decreasing evaporation, and supplying available N as it decomposes.

Just how an increase in organic matter content translates into improved water-supplying capacity is not well documented, although some measurements have been made. Wischmeier and Mannering (1965) showed that runoff decreased sharply as soil organic matter increased. The additional water entering the soil could make an important difference during periods of low rainfall and soil water deficits.

In western Nigeria, Lal et al. (1978) measured large increases in both infiltration rate and infiltration capacity where the primary crops were planted no-tillage into killed cover crops compared to conventional tillage without cover crops. They attributed much of the effects to high activity of earthworms (*Lumbricus* spp.) under the killed mulch. Average bulk density values of 1.35 Mg m^{-3} under the cover crops and 1.59 Mg m^{-3} for the conventionally tilled control amply supported their conclusion that cover crops and no-tillage improved soil structure. Soil water content, averaged over all cover crops, was nearly always as great or greater under the cover crop mulches than with the conventionally tilled control. Yields of the primary crops generally paralleled soil water storage; however, it is impossible to determine what caused the yield differences—greater water storage, mulch vs. no mulch, no-tillage vs. conventional tillage, or a combination of factors.

Evaporation Loss

A vegetative mulch retards water evaporation primarily in three ways: (i) by shading the soil from solar radiation, (ii) by insulating the soil from conduction of heat from the air, and (iii) by slowing water vapor movement from the soil to the air, thereby developing high humidity and permitting condensation within the mulch (Bond and Willis, 1969; Phillips, 1984).

A mulch slows the rate of evaporation loss but does not decrease the amount of water that ultimately can be lost from the soil. Therefore, a particular soil, whether mulched or bare, will lose equal amounts of water if given enough time. Clearly then, a mulch provides protection from short-term dry spells (7–14 d according to Bond and Willis, 1969), but not long-term droughts. Its greatest effectiveness occurs in the early growing season before a crop canopy forms and evaporation exceeds transpiration. However, Utomo's results (Table 8–6) showed that mulch from a hairy vetch cover crop conserved soil water in the surface 7.5 cm of a silt loam soil under no-tillage corn until well past canopy closure. Significant evaporation loss of soil water probably continued after canopy closure under conventional tillage but was decreased by the hairy vetch mulch with no-tillage.

Table 8-6. Soil water as affected by cover treatments and tillage during 1985 season. Adapted from Utomo (1986).

		Pre-plant		2 weeks		June		July		Sept.	
Depth, cm	Winter cover	CT†	NT	CT	NT	CT	NT	CT	NT	CT	NT
		soil water, kg kg^{-1}									
0 to 15	Fallow	0.24a*	0.23a	0.26b	0.27a	0.23b	0.24b	0.16c	0.18b	0.18c	0.20b
	Hairy vetch	0.20b	0.21b	0.24b	0.27a	0.21b	0.27a	0.15c	0.23a	0.18c	0.23a
15 to 30	Fallow	0.21a*	0.21a	0.25a	0.25a	0.23a	0.24a	0.18b	0.19b	0.19a	0.19a
	Hairy vetch	0.15b	0.15b	0.25a	0.24a	0.22b	0.24a	0.16c	0.22a	0.17a	0.20a
30 to 45	Fallow	0.22a*	0.23a	0.25a	0.26a	0.25a	0.25a	0.20a	0.20a		
	Hairy vetch	0.16b	0.15b	0.21b	0.22b	0.21b	0.22b	0.16b	0.19a		
45 to 60	Fallow	0.24a*	0.26a	0.27a	0.27a	0.27a	0.26a	0.23a	0.24a		
	Hairy vetch	0.19b	0.18b	0.19b	0.20b	0.22b	0.22b	0.19b	0.20b		

* Values for each depth increment and sampling date followed by different letters are significantly different ($P < 0.05$).
† CT = conventional tillage; NT = no-tillage.

Table 8-7. Mean soil water content and corn yield as affected by tillage and mulch. Adapted from Triplett et al. (1968).

Treatment	Available soil water†	Grain yield‡
	mm	Mg ha^{-1}
Plowed	42.7b*	5.97b
No-tillage, residue removed	43.7b	4.77c
No-tillage, normal residue	53.4ab	6.17ab
No-tillage, double residue	58.8a	6.62a

* Values followed by different letters are significantly different ($P < 0.05$).
† Average 2 yr, 15 June to 15 September; amount in 0- to 46-cm depth of soil.
‡ Average 2 yr.

Greater amounts of mulch, up to some optimal level, increase water conservation. Unger (1978) found that available soil water increased at the rate of 0.68 cm of water for each Mg ha^{-1} of mulch up to 12 Mg ha^{-1}. Triplett et al. (1968), working with 0, 1X, and 2X rates of mulch, found that the soil water benefits in no-tillage soil in comparison to plowed soil were closely related to the amount of mulch from crop residue (Table 8-7). The effect of the residue on corn yield paralleled the effect on available soil water with no-tillage. The higher yield with conventional tillage than with no-tillage without mulch (Table 8-7) is consistent with the usual observation that tillage increases N mineralization resulting in greater yields under similar soil water conditions.

Results by Corak (1987, unpublished data) indicate that, due to transpiration by hairy vetch, the soil had significantly ($P < 0.05$) less water at corn planting time under a hairy vetch cover crop than under corn residue fallow, e.g., 2.1 vs. 2.7 cm in the 0- to 7.5-cm depth. Within 28 d after planting corn, soil water content was significantly greater under the mulch. The difference disappeared a month later, and the values were virtually equal the remainder of the season.

Soil Water Depletion by Cover Crops

Along with conserving soil water, a cover crop depletes stored soil water before it is killed or plowed under, as shown by Corak's results above. Utomo (1986) determined water gravimetrically on soil samples taken at about 7-d intervals from the time corn was planted until near the end of July (Table 8-6). An additional sample was taken before harvest in September. A hairy vetch cover crop depleted the stored soil water to at least 60 cm; but, within about 2 weeks after it was killed at corn planting, the effect in the 0- to 15-cm soil depth of the no-tillage plots changed to water conservation. The water conservation effect continued in that layer of nontilled soil throughout the growing season; however, the depleted soil water was not restored below 30 cm during the entire season.

The depletion effect can produce serious problems with plant stands, seedling growth, and even crop yields during dry seasons. Furthermore, it may alter the amount of water used by the summer crop by decreasing the amount of plant-available soil water. The clear indication that soil water removed by the cover crop may not be restored during the season below about 30 cm is ample evidence to support this hypothesis. However, the net effect of a cover crop on soil water storage depends on the amount and timing of rainfall and the amount of infiltration vs. evaporation and transpiration losses during both the cover crop and summer crop periods.

NITROGEN FROM ANNUAL LEGUME COVER CROPS

Cover Crop Yield and Nitrogen Content

Dry matter yields and N concentration provide a measurement of the N content of the aboveground portion of a cover crop. The value represents

Table 8-8. Dry matter production and N content of legume cover crops.

Cover crop	Dry matter†	Nitrogen content	Reference and notes
	Mg ha^{-1}	kg ha^{-1}	
Hairy vetch	5.1a*	209a	Ebelhar et al. (1984), Lexington, KY; avg. 1980–1981
Bigflower vetch	1.9d	60b	
Crimson clover	2.4c	56b	
Rye	3.4b	36c	
Hairy vetch	3.3a*	130a	Utomo (1986), Lexington, KY; avg. 1984–1985
Bigflower vetch	2.5b	76b	
Rye	3.4a	40c	
Hairy vetch	3.5	122	Frye et al. (1986), Princeton, KY; avg. 1981–1986
Bigflower vetch	2.9	99	
Rye	2.3	21	
Hairy vetch	3.5‡	136‡	Neely et al. (1987), Georgia; avg. 1985–1986
Crimson clover	4.0‡	111‡	
Berseem clover (*Trifolium alexandrinum*)	1.5	45	
Austrian winter pea	1.6	68	
Wheat	1.5	29	
Hairy vetch	4.2	153	Hargrove (1986), Georgia; avg. 1981–1983
Common vetch	4.3	134	
Crimson clover	7.2	170	
Subterranean clover	4.0	114	
Rye	4.0	38	
Hairy vetch	4.6	130	Elliott et al., (1987), Washington; values for 1985
Austrian winter pea	6.4	137	
Hairy vetch	3.7	158	Hoyt (1987), North Carolina; avg. 1982–1985
Crimson clover	4.7	129	
Austrian winter pea	4.6	161	
Rye	6.3	100	

* Values followed by different letters within an experiment are significantly different ($P < 0.05$). Absence of letters indicates statistical data unavailable.
† Aboveground portion sampled in spring, except Elliott et al. sampled in summer.
‡ Average of two location.

N uptake from residual soil inorganic nitrogen and mineralized soil organic nitrogen, as well as biologically fixed N.

Table 8–8 shows cover crop yields and N content values from several published studies. Several points can be made from these data. Hairy vetch is superior in N content to the other legume cover crops in most studies. Dry matter production by crimson clover is greatest in the southern part of the southeastern USA, whereas hairy vetch produces more dry matter in the northern part of the region. Most studies included a small grain cover crop, which allows one to estimate the amount of N fixed by the legume cover crop. If one assumes that the legume and small grain take up equal amounts of mineralized N from the soil, the difference between the N contents of the two cover crops approximates N_2 fixation. The range was from 20 to 173 kg of N ha^{-1}, with hairy vetch excelling in most experiments.

Nitrogen Uptake by Summer Crops

The increased amount of N from a cover crop results in greater N availability to summer crops. Table 8–9 shows results obtained in Kentucky by Utomo (1986) and Ebelhar et al. (1984). When N fertilizer was applied at limiting rates, i.e., all cases except with 170 kg of N ha^{-1} in Utomo's study, the greater amount of available N from hairy vetch resulted in a singificantly ($P < 0.05$) higher N concentration in the corn plant tissue.

In a study with ^{15}N-labeled hairy vetch and rye, Varco (1986) found that N uptake by corn was equal on rye plots and fallow plots, but was about 60 kg ha^{-1} greater with the hairy vetch cover crop (Fig. 8–1). The N uptake was greatly enhanced by plowing compared to no-tillage. This effect was attributed to greater mineralization of both hairy vetch N and soil N where the soil was plowed.

Table 8–9. Nitrogen concentration in whole corn plants following legume winter cover crops.

Winter cover	Tillage†	Fertilizer N, kg ha^{-1}: 0	170	Reference and notes
		N, g kg^{-1}		
Fallow	CT	13e*	23b	Utomo (1986), Lexington, KY; sampled at silking.
	NT	17d	24ab	
Hairy vetch	CT	21c	26ab	
	NT	21c	27a	
Rye	CT	14e	24ab	
	NT	14e	24ab	
		0	100	
Fallow	NT	25b*	29b	Ebelhar et al. (1984), Lexington, KY; sampled at 8- to to 10-leaf stage.
Hairy vetch	NT	27a	34a	
Bigflower vetch	NT	25b	30b	
Crimson clover	NT	24bc	30b	
Rye	NT	22c	31b	

* Different letters within each experiment indicate $P < 0.05$.
† CT = conventional tillage; NT = no-tillage.

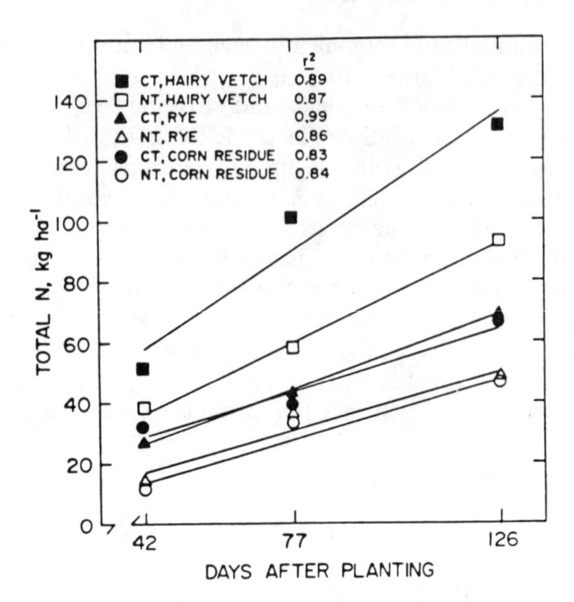

Fig. 8-1. Total N uptake by corn as affected by winter cover treatment and tillage in 1984.

The total N taken up by corn in Varco's experiment included both N mineralized from hairy vetch and N from the soil. So, another of Varco's objectives was to determine the amount of N uptake contributed by the ^{15}N-labeled hairy vetch. About 35 kg ha^{-1} (32%) of hairy vetch N was removed by the corn in the plowed treatment in both 1984 and 1985 (Fig. 8-2). In contrast, about 22 kg of N ha^{-1}, or 20%, was obtained from hairy vetch under no-tillage conditions. The difference due to tillage was apparent at the first sampling date (42 d after planting) and continued until maturity of the corn at 126 d. Plowing increased the amount of N obtained from hairy vetch and rye mulch by 12 and 10%, respectively.

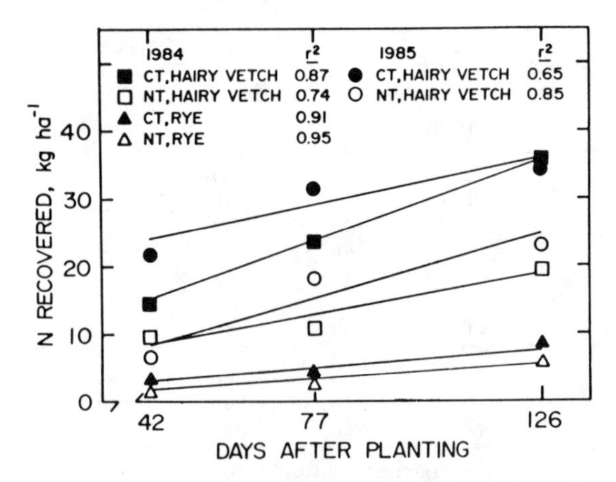

Fig. 8-2. Recovery of N by corn from ^{15}N-labeled cover crops.

Effect of Cover Crops on Soil Nitrogen

Legume cover crops increase both plant-available inorganic soil nitrogen and organic soil nitrogen. Nitrogen from cover crop residue not recovered by the first summer crop may be immobilized as organic soil nitrogen and become available to succeeding crops. Total N uptake by corn from mineralization of N from hairy vetch cover crops of previous years was about 30 kg of N ha^{-1} with conventional tillage and about 35 kg of N ha^{-1} with no-tillage in Varco's experiment (1986), which had a history of several years of hairy vetch cover crop and no-tillage corn.

Inorganic Soil Nitrogen

Ebelhar et al. (1984) measured KCl extractable NH_4 and NO_3 (inorganic nitrogen) in the soil during the corn-growing season following winter cover treatments of hairy vetch, rye, and fallow (corn residue). There was no difference among the treatments at corn planting, but 3 to 5 weeks later the hairy vetch plots were significantly ($P < 0.05$) higher in inorganic nitrogen than rye and fallow plots, which were virtually equal. In the 0- to 7.5-cm soil depth, the additional inorganic soil nitrogen with hairy vetch compared to fallow was about 22 and 30 mg of N kg^{-1} soil, respectively, with 0 and 100 kg ha^{-1} fertilizer N. By the end of July, the effect of the hairy vetch had nearly disappeared.

Utomo (1986) measured inorganic soil nitrogen as affected by soil depth, winter cover treatment, N-fertilizer rate, and tillage system during 1984 and 1985 (Table 8–10). As with the results of Ebelhar et al. (1984), there were no significant differences among treatments at the time the cover crops were plowed under or killed. Results with rye were virtually identical to those with

Table 8–10. Soil inorganic nitrogen (KCl extractable NH_4 and NO_3) as affected by winter cover, tillage system, and N fertilizer rate at 4 and 8 weeks after corn planting and N application in 1984. Adapted from Utomo (1986).

Sampling date†	Depth	Winter cover	Fertilizer N, kg ha^{-1}: 0, CT‡	0, NT	170, CT	170, NT
	cm		N, mg kg^{-1}			
4 weeks	0–7.5	Fallow	16e*	17d	110b	108b
		Hairy vetch	69c	39de	178a	73c
4 weeks	7.5–15	Fallow	13d	10d	44b	37bc
		Hairy vetch	57b	22cd	86a	52b
8 weeks	0–7.5	Fallow	16c	15c	33b	30b
		Hairy vetch	15c	15c	50a	41ab
8 weeks	7.5–15	Fallow	10d	12d	17cd	24bc
		Hairy vetch	14d	11d	56a	28b

* Values for each sampling date and soil depth followed by different letters are significantly different ($P < 0.05$).
† Time after planting.
‡ CT = conventional tillage; NT = no-tillage.

fallow and the data were omitted from Table 8-10. Thirty days after being plowed under or killed, hairy vetch resulted in a substantial increase in soil inorganic nitrogen, especially with plowing. With no-tillage, the increase was generally nonsignificant ($P > 0.05$). The greatest differences occurred between the 0 and 170 kg ha^{-1} fertilizer N rates. These differences lasted throughout most of the growing season. The largest amount of inorganic nitrogen at each sampling date was where hairy vetch was plowed under and 170 kg ha^{-1} fertilizer N applied.

Utomo's results agreed with those obtained by Varco (1986) using ^{15}N techniques. Plowing greatly enhanced mineralization of N from hairy vetch, but this did not result in an increase in corn yield, except at the zero N-fertilizer rate.

Organic Soil Nitrogen

Legume winter cover crops increased both the soil organic carbon and organic nitrogen relative to fallow and nonlegume winter cover crops in Kentucky and Georgia (Table 8-11). The relative increase by hairy vetch was considerably greater in the 0- to 7.5-cm depth of soil with no-tillage than with conventional tillage (Utomo, 1986). This probably accounts for the slower release of N and less N recovered in no-tillage reported by Varco (1986) (Fig. 8-1 and 8-2) and the observation of Ebelhar et al. (1984) that N deficiency in corn appeared last of all treatments in the hairy vetch plots. The additional N retained in the soil by maintaining a higher organic matter content with legume cover crops and no-tillage is rather substantial. Estimated

Table 8-11. Soil organic carbon and nitrogen in 0- to 7.5-cm depth as affected by legume winter cover crops.

Winter cover	Tillage†	Organic carbon	Organic nitrogen	Reference and notes
		— g kg^{-1} —		
Fallow	CT	16.8c*	1.64d	Utomo (1986), Lexington, KY;
	NT	23.7b	2.14b	CT = 7 yr no-tillage and 2 yr
Hairy vetch	CT	16.9c	1.78c	plowed and disked, NT = 9 yr
	NT	25.2a	2.38a	no-tillage. Summer crop was
Rye	CT	15.2c	1.64d	continuous corn, 170 kg ha^{-1}
	NT	22.4b	2.19b	fertilizer N.
Fallow	NT	10.6b*	1.2b	Frye et al. (1986), Princeton, KY;
Hairy vetch	NT	13.5a	1.5a	0- to 7.5-cm soil depth after 6 yr
Bigflower vetch	NT	12.7a	1.4a	no-tillage corn; avg. of three
Rye	NT	11.5b	1.2b	fertilizer N rates.
Fallow	NT	7.9b*	0.58c	Hargrove (1986), Georgia; after 3 yr
Hairy vetch	NT	9.7a	0.80ab	no-tillage grain sorghum; avg. of
Crimson clover	NT	8.4b	0.65bc	four N fertilizer rates.
Subterranean clover	NT	10.0a	0.81a	
Common vetch	NT	10.2a	0.63c	
Rye	NT	8.7b	0.65bc	

* Values followed by different letters for each experiment are significantly different ($P < 0.05$).

† CT = conventional tillage; NT = no-tillage.

values for the 0- to 7.5-cm layer of soil based on Utomo's organic soil nitrogen data at the end of the growing season are shown in Table 8-12.

CROP YIELDS IN RESPONSE TO LEGUME COVER CROPS

A number of studies have shown that legume winter cover crops increase yields of summer crops (Table 8-13). The yield response to legume cover crops is greatest at low or zero N rates, as would be expected. However, surprisingly high yield increases due to legume cover crops have been obtained with rather high rates of N in some studies. For example, the highest corn yield in the study by Utomo (1986) occurred with 170 kg ha^{-1} fertilizer N on the hairy vetch/no-tillage plots. Ebelhar et al. (1984) found similar results with 100 kg ha^{-1} fertilizer N and hairy vetch, as did Decker et al. (1987) with 80 kg of N ha^{-1} and hairy vetch. Thus, legume cover crops appear to increase crop yields over a wide range of N-fertilizer rates.

Type and Implication of Yield Responses

Grain-yield responses to N fertilizer with and without legume cover crops generally fit three hypothetical patterns, according to Smith et al. (1987). Type A curve may or may not show a response to the legume cover crop (beyond the response to N fertilizer) at zero, low, and perhaps medium N-fertilizer rates, but yields with both the legume and N fertilizer are equal at higher N rates. Yields shown in Table 8-13 that match this pattern are those for grain sorghum obtained by Hargrove (1986) and probably Touchton et al. (1982), although they did not include a fallow treatment in their experiment for comparison. This type of response implies that additional N is the only significant effect of the legume on crop yields (Smith et al., 1987). In some cases, such as Touchton et al. (1982) and Hargrove (1986), the cover crop supplies all of the N needs of the summer crop. It appears that grain sorghum may typically respond in this way.

Type B curve is a rather typical yield response to N fertilizer both with and without a legume cover crop; but throughout the range of fertilizer rates,

Table 8-12. Soil organic nitrogen in 0 to 7.5 cm of soil in September 1985 as affected by winter cover, tillage, and N fertilizer. Adapted from Utomo (1986).

	Fertilizer N, kg ha^{-1}			
	0		170	
Winter cover	CT†	NT	CT	NT
	N,‡ kg ha^{-1}			
Fallow	1580f*	1920c	1640ef	2140b
Hairy vetch	1760de	2180b	1780d	2380a
Rye	1650e	1910c	1640ef	2190b

* Values followed by different letters are significantly different ($P < 0.05$).
† CT = 7 yr no-tillage and 2 yr conventional tillage; NT = 9 yr no-tillage.
‡ Values based on assumption of similar soil bulk densities at sampling in September.

Table 8-13. Yields of summer crops following legume winter cover crops.

Summer crop	Winter cover	Tillage†	Fertilizer N, kg ha⁻¹ 0	85	170	Reference and notes
			Yields, Mg ha⁻¹			
Corn	Fallow	CT	4.0	5.9	6.6	Utomo (1986), Lexington, KY; avg. 1984-1985.
		NT	2.8	5.7	6.7	
	Hairy vetch	CT	7.1	6.9	7.7	
		NT	6.4	7.2	8.0	
			0	50/ 85‡	100/ 170‡	
Corn	Fallow		1.1	2.7	5.2	Herbek et al. (1987), Princeton, KY; avg. 1980-1986.
	Hairy vetch		3.1	4.3	6.0	
	Bigflower vetch		2.8	4.4	6.2	
			0	50	100	
Corn	Fallow		3.8b*	5.2b	6.8bc	Ebelhar et al. (1984), Lexington, KY; avg. of 1977-1981.
	Hairy vetch		6.4a	6.9a	9.1a	
	Bigflower vetch		4.2b	6.6a	6.6c	
	Crimson clover		4.4b	5.6b	7.4bc	
			0	56	112	
Corn	Fallow		4.0	6.4	7.0	Mitchell and Teel (1977), Delaware; avg. 1974-1975.
	Rye and hairy vetch		6.1	7.2	8.1	
	Rye		4.0	5.8	7.3	
			0	90	180	
Corn	Fallow	CT	1.6	4.7	4.8	Adams et al. (1970), Georgia; avg. 1958-1964.
	Hairy vetch	CT	5.3	5.6	5.5	
			0		112	
Grain sorghum	Fallow		2.9b		3.9a	Hargrove (1986), Georgia; avg. of 1981-1983.
	Hairy vetch		4.0a		3.8a	
	Common vetch		3.7a		3.9a	
	Crimson clover		3.9a		4.2a	
	Subterranean cl.		3.9a		3.9a	
			0	45	90	
Grain sorghum	Crimson clover		5.2	5.3	5.4	Touchton et al., (1982), Georgia; avg. 1978-1980.
			0	34	68	
Cotton	Fallow		0.59	0.75	0.81	Touchton et al. (1984), Alabama; avg. 1981-1982.
	Common vetch		0.83	0.92	0.90	
	Crimson clover		0.89	0.81	0.72	

* Values followed by different letters within an experiment are significantly different ($P < 0.05$). Absence of letters indicates statistical data unavailable.
† No-tillage (NT) except where indicated conventional tillage (CT).
‡ 50 and 100 kg of N ha⁻¹ during 1980 to 1983; 85 and 170 during 1984 to 1986.

the yields with the legume remain roughly parallel and above those with N fertilizer alone. Yield results that fit this pattern include those obtained by Herbek et al. (1987), Utomo (1986), Ebelhar et al. (1984), Mitchell and Teel

(1977) and Adams et al. (1970). This appears to be the usual response of corn to legume cover crops, at least at low and moderate rates of N fertilizer. This type of response suggests that the legume cover crop provides yield-influencing benefits in addition to N. It is not known what these benefits are. Speculation has centered around such things as increased soil water efficiency, improved weed control, and the nebulous "rotation effect."

In type C response, yields with a legume exceeds those without a legume at zero or low N-fertilizer rates, but with sufficient N supplied by fertilizer, yields without a legume are superior to those with a legume. The implication is that, in addition to the benefit of N contribution where N is limiting, the legume cover crop produces a negative effect on yields. Cotton following crimson clover responded in this manner in the results shown by Touchton et al. (1984) (Table 8-13), which they attributed to delayed maturity caused by excess N. One of the best examples of type C response, however, is probably that obtained with no-tillage corn by Corak in Kentucky (Smith et al., 1987). Corn yields at 255 kg ha^{-1} of fertilizer N decreased with both the 1X and 2X rates of hairy vetch mulch compared to treatments where the hairy vetch was cut and removed.

The different types of yield response of summer crops to legume winter cover crops indicate that in some cases the cover crops provide benefits beyond their N contribution, in other cases they do not, and, in a few situations, they produce dominating negative effects and are an overall liability. Yields may respond one way one year and another way the next year, depending on the conditions. If N contribution is the predominate effect, it can be negated entirely by application of N fertilizer. If there are benefits beyond N contribution, yields with a legume cover crop will exceed those with the fallow treatment regardless of N-fertilizer rates. Where the combined detrimental effects exceed the combined beneficial effects of the legume cover crop, a negative yield response will occur.

Increased Yields Over Time

In addition to annual benefits, yield benefits may accrue over time. Frye et al. (1985b) pointed out that, on a year-by-year basis, data from Ebelhar et al. (1984) showed that the gap between corn grain yields with the hairy vetch cover crop and the fallow treatments widened over a 5-yr period. For example, the difference in yields between fallow and hairy vetch treatments each with 100 kg ha^{-1} of fertilizer N increased by an average of 0.51 Mg ha^{-1} yr^{-1}. Frye et al. (1985b) labeled this effect "increased soil productivity" and attributed it mainly to the increased N supply as organic nitrogen was built up in the soil, but stated that it may have been partially due to improved physical condition and water relations resulting from increased soil organic matter.

Nitrogen Equivalence of Legume Cover Crops

The yield response to a legume cover crop has been used by some researchers to estimate the amount of N provided to a summer crop by the

Table 8-14. Estimated fertilizer N equivalence of legume winter cover crops.

Summer crop	Cover crop		Fertilizer N equivalence†	References and notes
			kg ha^{-1}	
Corn	Hairy vetch		87	Ebelhar et al. (1984), Lexington, KY; avg. of 5 yr.
	Bigflower vetch		15	
	Crimson clover		22	
Corn	Hairy vetch		66	Herbek et al. (1987), Princeton, KY; avg. of 6 yr.
	Bigflower vetch		64	
Corn	Hairy vetch	CT	203	Utomo (1986), Lexington, KY; avg. of 2 yr; CT = conventional tillage, NT = no-tillage.
		NT	103	
	Bigflower vetch	CT	101	
		NT	54	
Corn	Hairy vetch-Oats		163	Mitchell and Teel (1977), Delaware; avg. of 2 yr; corn irrigated.
	Hairy vetch-Oats		110	
	Hairy vetch-Rye		110	
	Hairy vetch-Rye		73‡	
Grain sorghum	Hairy vetch		97	Hargrove (1986), Georgia; avg. of 3 yr
	Common vetch		61	
	Crimson clover		92	
	Subterranean cl.		61	
Grain sorghum	Hairy vetch		61	Herbek et al. (1987),Princeton, KY; avg. of 2 yr.
	Bigflower vetch		75	
Cotton	Hairy vetch		68	Touchton et al. (1984), Alabama.
	Crimson clover		68	
Spring barley (*Hordeum vulgare* L.)	Hairy vetch (green-manured)		75	Meyer (1987), North Dakota; avg. 1985–1986, one location 1985, two locations 1986.
	Hairy vetch (green-manured)		149‡	

† Cover crop treatment without N fertilizer compared to winter fallow treatment with N fertilizer.

‡ Legume cover crop treatment without N fertilizer compared to nonlegume cover crop treatment with fertilizer.

legume cover crop. The amount of N fertilizer required in a fallow or small grain cover crop treatment to give a yield equal to that obtained on the legume cover crop treatment without N fertilizer is termed the fertilizer-N equivalence of the legume crop (Hargrove, 1986; Smith et al., 1987).

Table 8-14 shows the fertilizer-N equivalence values from several experiments. Other values were listed by Smith et al. (1987). Hairy vetch was the most effective cover crop, with values ranging from about 60 kg of N ha^{-1} with grain sorghum (Herbek et al., 1987) to about 200 kg of N ha^{-1} with conventionally tilled corn (Utomo, 1986). Crimson clover was about equal to hairy vetch in Georgia (Hargrove, 1986) and Alabama (Touchton et al., 1984). Utomo's results show a greater fertilizer-N equivalence in plowed soil than in no-tillage soil, as would be expected considering the greater N mineralization with plowing.

There are inherent problems with the N-equivalence technique for estimating the N contribution of a winter legume cover crop to a summer crop. Hargrove (1986) pointed out that this approach presumes that N contribution is the main way in which the legume cover crop affects yields and ignores the effects on soil temperature, soil water relations, soil structure, and other factors and their subsequent effect on yields. He suggested that using N content of the grain might be a more appropriate technique, since it more often correlates with available N than yield does.

Smith et al. (1987) pointed out that a major advantage of the fertilizer-N equivalence technique is that the value of the legume cover crop can be estimated, for management purposes, in terms of two easily determined parameters—crop yield and N-fertilizer applied. However, they expressed some serious concerns about the validity of the technique for estimating the amount of N provided to the summer crop, the effect of the legume cover crop on N accumulation by a summer crop, the amount that N fertilizer can be decreased following a legume cover crop, or the actual amount of N released from the legume. They suggested that a more suitable estimate of the N-fertilizer replacement value of a legume cover crop might be the difference in the amount of N fertilizer required to attain optimum yield or optimum N uptake with and without a legume cover crop. They identified the greatest problem with this approach as the lack of adequate data from past experiments.

In a practical context, what is needed is a means for estimating the N-fertilizer requirement for a summer crop following a legume winter cover crop. Since there are both beneficial and detrimental effects of a legume cover crop, having a good estimate of the N contribution does not necessarily answer the question. An excellent example can be found in the data of Ebelhar et al. (1984). The calculated fertilizer-N equivalence value for hairy vetch was 87 kg of N ha^{-1} (Table 8–14), but by far the highest net economic return was from corn grown with the hairy vetch plus 100 kg ha^{-1} fertilizer N (Frye et al., 1985b). To suggest that farmers could decrease their fertilizer N by 87 kg ha^{-1} from a normally adequate rate for corn following hairy vetch would be an extremely costly recommendation under those conditions. Similar results would be expected anywhere that benefits beyond N contribution are acquired from a legume cover crop.

Certainly, then, fertilizer-N equivalence has weaknesses in certain cases for estimating the N-fertilizer replacement value of legume cover crops. The technique would be satisfactory in case of a type A yield response, i.e., benefit from legume is mainly from N contribution, but unsatisfactory with a type B yield response where there are significant benefits from the legume beyond its N contribution. Most type C yield responses are probably caused by either too heavy mulch or too much N. Since the N equivalence value is based on both, it could be a useful tool to predict when a legume cover crop might decrease crop yield. In such a case, a farmer could kill the cover crop early or remove some of it before planting the summer crop and avoid the possibility of decreased yield.

SUMMARY AND CONCLUSION

Annual legume winter cover crops can conserve water and supply a substantial part of the N needed for optimum yields of nonleguminous summer crops. Several legume species are adaptable in such cropping strategies, but hairy vetch appears to perform best in the mid-section of the USA from Delaware, Maryland, Kentucky, to Nebraska, while crimson clover as well as hairy vetch appears to be well adapted to the southern part of the Southeast.

The cover crops can be established by surface interseeding into a summer crop just before leaf drop, or they can be planted after the summer crop has been harvested. A cover crop may be plowed under as a green manure or chemically killed to form a mulch for no-tillage or another form of conservation tillage. Plowing the legume under enhances N recovery by the summer crop, but killing it for a surface mulch enhances its soil and water conservation value.

Benefits from a legume winter cover crop include the following:

- Provides biologically fixed N to succeeding summer crops. The amount of N provided varies with soil, climate, and management conditions. The fertilizer N equivalence may be a satisfactory technique for some crops and situations but not for others. It is suggested that a recommendation for the amount of N fertilizer to be used following a legume cover crop should be based on the estimated amount needed to attain the expected optimum economic yield.
- Increases soil organic carbon and organic nitrogen relative to winter fallow. The increase is greater under no-tillage than under conventional tillage.
- Improves soil structure, increases water infiltration, decreases runoff, and decreases evaporation losses of water if left as a surface mulch.
- Enhances yields of summer crops, often beyond what can be obtained with N fertilizer alone. This emphasizes the benefits other than the N contribution, such as, increased soil water conservation, improved weed control, and other less obvious benefits.
- Increases soil productivity. One study showed that yields of no-tillage corn following hairy vetch increased at an annual rate of 0.51 Mg ha^{-1} relative to the fallow treatment.

The two most important liabilities of using legume winter cover crops are depletion of stored soil water in the early spring and the cost of seeding.

A vigorously growing legume cover crop definitely decreases the amount of stored soil water to a depth of 60 cm or more. Early killing or plow-down can lessen this effect, with the sacrifice of some N and mulch value. Within 2 weeks after kill-down in no-tillage the legume mulch clearly becomes effective in soil water conservation, provided rainfall occurs or irrigation is applied. However, with normal rainfall, soil water does not appear to be restored below about 30 cm. The decrease in soil water content may be a distinct advantage in the case of wet soils.

Seeding costs can be decreased somewhat by using a self-reseeding management technique. Crimson clover is well-adapted to this approach in the southern portion of the Southeast while bigflower vetch has excellent prospects for this type of management farther north. Unpublished work by

W.L. Hargrove in Georgia (1987, personal communication) indicates that seeding rates for some legumes may be decreased somewhat from recommended rates without decreasing their effectiveness as cover crops. This could also lower seeding costs.

Using legume winter cover crops to supplement N fertilizer, conserve rainfall or irrigation water after kill-down, and help control soil erosion is indisputably feasible from an agronomic viewpoint. Under present price conditions, the economic feasibility of the practice is doubtful if the sole purpose or benefit of the cover crop is its N-fertilizer replacement value, because this may not offset the seeding costs (Allison and Ott, 1987; Ott, 1987). However, if yields of summer crops are increased significantly beyond that attributable to the N contribution, the legume cover crop will be profitable (Frye et al., 1985b; Frye, 1986).

Increases in grain yields and soil improvement that have occurred in some studies make using legume winter cover crops a potentially important cropping strategy for the future, as in the past. Perhaps legume cover crops are the key to long-term, sustainable, intensive row-crop agriculture. They also hold the promise of restoring at least some of the soil productivity lost due to our exploitative soil management practices of the past (Langdale et al., 1987).

Research Needs

Additional research is needed to achieve the potential benefits of legume cover crops. Future research on cover crops should address the following objectives:

- Develop improved cover crop legumes. This is perhaps one of the most critical needs. It should include breeding or genetically engineering new lines as well as continuing to test and improve the adaptation of existing species and lines as cover crops through selection and management practices. Allied with this is the possibility of developing more efficient strains of rhizobia bacteria for either old or new lines of legumes.
- Extend the range of adaptability of winter legume cover crops northward. Several species perform excellently in the Southeast, but farther north, winter hardiness and the capacity to break dormancy early in the spring and produce large quantities of dry matter and N before corn planting time are lacking in most species.
- Develop cropping strategies and management schemes to make cover crops more compatible in common crop rotations.
- Continue research on legume N transformations and plant availability in the soil, as well as its ultimate fate. The purpose should be to improve the efficiency of legume cover crops as a source of N for nonlegume crops.
- Improve techniques for making N-fertilizer recommendations for summer crops based on N contribution and other benefits derived from legume cover crops.

- Decrease seeding costs of legume cover crops. Two approaches currently being researched are to (i) improve self-reseeding technology and (ii) lower seeding rates of legumes without decreasing their effectiveness as a cover crop. Other possible approaches need to be studied.
- Determine the potential value of legume cover crops in restoring the crop productivity of eroded or degraded soils.

An awareness that the value of legume cover crops has the potential to go far beyond their N contribution has rekindled interest in adapting them to modern cropping strategies. Agriculturalists should foster that interest and develop that potential for a future agriculture that is agronomically, economically, and environmentally sound.

REFERENCES

Adams, W.E., H.D. Morris, and R.N. Dawson. 1970. Effect of cropping systems and nitrogen levels on corn (*Zea mays*) yields in the Southern Piedmont Region. Agron. J. 62:655–659.

Allison, J.R., and S.L. Ott. 1987. Economics of using legumes as a nitrogen source in conservation tillage systems. p. 145–150. *In* J.F. Power (ed.) The role of legumes in conservation tillage systems. Soil Conserv. Soc. Am., Ankeny, IA.

Auld, D.L., B.L. Bettis, M.J. Dial, and G.A. Murray. 1982. Austrian winter and spring peas as green manure crops in northern Idaho. Agron. J. 74:1047–1050.

Beale, O.W., G.B. Nutt, and T.C. Peele. 1955. The effects of mulch tillage on runoff, erosion, soil properties, and crop yields. Soil Sci. Soc. Am. Proc. 19:244–247.

Blevins, R.L., G.W. Thomas, M.S. Smith, W.W. Frye, and P.L. Cornelius. 1983. Changes in soil properties after 10 years continuous non-tilled and conventionally tilled corn. Soil Tillage Res. 3:135–146.

Bond, J.J., and W.O. Willis. 1969. Soil water evaporation: Surface residue rate and placement effects. Soil Sci. Soc. Am. Proc. 33:445–448.

Brown, S.M., T. Whitwell, J.T. Touchton, and C.H. Burmester. 1985. Conservation tillage systems for cotton production. Soil Sci. Soc. Am. J. 49:1256–1260.

Dabney, S.M., and J.L. Griffin. 1987. Efficacy of burndown herbicides on winter legume cover crops. p. 122–124. *In* J.F. Power (ed.) The role of legumes in conservation tillage systems. Soil Conserv. Soc. Am., Ankeny, IA.

Decker, A.M., J.F. Holderbaum, R.F. Mulford, J.J. Meisinger, and L.R. Vough. 1987. Fall-seeded legume nitrogen contributions to no-till corn production. p. 21–22. *In* J.F. Power (ed.) The role of legumes in conservation tillage systems. Soil Conserv. Soc. Am., Ankeny, IA.

Ebelhar, S.A., W.W. Frye, and R.L. Blevins. 1984. Nitrogen from legume cover crops for no-tillage corn. Agron. J. 76:51–55.

Elliott, L.F., R.I. Papendick, and D.F. Bezdicek. 1987. Cropping practices using legumes with conservation tillage and soil benefits. p. 81–89. *In* J.F. Power (ed.) The role of legumes in conservation tillage systems. Soil Conserv. Soc. Am., Ankeny, IA.

Felton, W.L., D.M. Freebairn, N.A. Fettell, and J.B. Thomas. 1987. Crop residue management. p. 171–193. *In* P.S. Cornish and J.E. Pratley (ed.) Tillage: New directions in Australian agriculture. Inkata Press, Melbourne, Australia.

Frye, W.W. 1986. Economics of legume cover crops. Soil science news and views. Univ. of Kentucky, Lexington, KY.

----, J.H. Herbeck, and R.L. Blevins. 1983. Legume cover crops in production of no-tillage corn. p. 179–191. *In* W. Lockeretz (ed.) Environmentally sound agriculture. Prager Publ., New York.

----, ----, and ----. 1986. Nitrogen from legume cover crops for no-tillage corn. 1986. Grains research. Univ. of Kentucky, Lexington.

----, ----, ----, and J.J. Varco. 1984. Nitrogen from legume cover crops for no-tillage corn. 1984. Agronomy research report. Progress Rep. 281. Univ. of Kentucky, Lexington.

----, O.L. Bennett, and G.J. Buntley. 1985a. Restoration of crop productivity on eroded or degraded soils. p. 335–356. *In* R.F. Follett and B.A. Stewart (ed.) Soil erosion and crop productivity. ASA, Madison, WI.

----, W.G. Smith, and R.J. Williams. 1985b. Economics of winter cover crops as a source of nitrogen for no-till corn. J. Soil Water Conserv. 40:246–249.

Hale, J.K., N.L. Hartwig, and L.D. Hoffman. 1984. Cyanazine losses in runoff from no-tillage corn in "living" and dead mulches vs. unmulched conventional tillage. J. Environ. Qual. 13:105–110.

Hargrove, W.L. 1982. Proceedings of the minisymposium on legume cover crops for conservation tillage production systems. Spec. Publ. 19. Univ. of Georgia College of Agriculture. Univ. of Georgia, Athens.

----. 1986. Winter legumes as a nitrogen source for no-till grain sorghum. Agron. J. 78:70–74.

----, and W.W. Frye. 1987. The need for legume cover crops in conservation tillage production. p. 1–5. *In* J.F. Power (ed.) The role of legumes in conservation tillage systems. Soil Conserv. Soc. Am., Ankeny, IA.

Herbek, J.H., W.W. Frye, and R.L. Blevins. 1987. Nitrogen from legume cover crops for no-till corn and grain sorghum. p. 51–52. *In* J.F. Power (ed.) The role of legumes in conservation tillage systems. Soil Conserv. Soc. Am., Ankeny, IA.

Holderbaum, J.F., A.M. Decker, F.R. Mulford, J.J. Meisinger, and L.R. Vough. 1987. Forage contributions of winter legume cover crops in no-till corn production. p. 98–99 *In* J.F. Power (ed.) The role of legumes in conservation tillage systems. Soil Conserv. Soc. Am., Ankeny, IA.

Hoyt, G.D. 1987. Legumes as green manure in conservation tillage. p. 96–98. *In* J.F. Power (ed.) The role of legumes in conservation tillage systems. Soil Conserv. Soc. Am., Ankeny, IA.

----, and W.L. Hargrove. 1986. Legume cover crops for improving crop and soil management in the southern United States. HortScience 21:397–402.

Jauregui, H.E.M. 1986. Management of perennial and reseeding legume cover crops for no-tillage corn production. M.S. thesis. Univ. of Kentucky, Lexington.

Lal, R., G.F. Wilson, and B.N. Okigbo. 1978. No-till farming after various grasses and leguminous cover crops in tropical Alfisol. I. Crop performance. Field Crops Res. 1:71–84.

Langdale, G.W., R.R. Bruce, and A.W. Thomas. 1987. Restoration of eroded Southern Piedmont land in conservation tillage systems. p. 142–143. *In* J.F. Power (ed.) The role of legumes in conservation tillage systems. Soil Conserv. Soc. Am., Ankeny, IA.

Meyer, D.W. 1987. Influence of green-manure, hayed, or grain legumes on grain yield and quality of the following barley crop in the northern Great Plains. p. 94–95. *In* J.F. Power (ed.) The role of legumes in conservation tillage systems. Soil Conserv. Soc. Am., Ankeny, IA.

Mitchell, W.H., and M.R. Teel. 1977. Winter annual cover crops for no-tillage corn production. Agron. J. 69:569–573.

Nanni, C., and C.S. Baldwin. 1987. Interseeding in corn. p. 26–27. *In* J.F. Power (ed.) The role of legumes in conservation tillage systems. Soil Conserv. Soc. Am., Ankeny, IA.

Neely, C.L., K.A. McVay, and W.L. Hargrove. 1987. Nitrogen contribution of winter legumes to no-till corn and grain sorghum. p. 48–49. *In* J.F. Power (ed.) The role of legumes in conservation tillage systems. Soil Conserv. Soc. Am., Ankeny, IA.

Ngalla, C.F., and D.J. Eckert. 1987. Wheat-red clover interseeding as a nitrogen source for no-till corn. p. 47–48. *In* J.F. Power (ed.) The role of legumes in conservation tillage systems. Soil Conserv. Soc. Am., Ankeny, IA.

Ott, S.L. 1987. An economic and energy analysis of crimson clover as a nitrogen fertilizer substitute in grain sorghum production. p. 150–151. *In* J.F. Power (ed.) The role of legumes in conservation tillage systems. Soil Conserv. Soc. Am., Ankeny, IA.

Oyer, L.J., and J.T. Touchton. 1987. Nitrogen fertilizer requirements for corn as affected by legume cropping systems and rotations. p. 44–45. *In* J.F. Power (ed.) The role of legumes in conservation tillage systems. Soil Conserv. Soc. Am., Ankeny, IA.

Phillips, R.E. 1984. Soil moisture. p. 66–86. *In* R.E. Phillips and S.H. Phillips (ed.) No-tillage agriculture: Principles and practices. Van Nostrand Reinhold Co., New York.

Smith, M.S., W.W. Frye, and J.J. Varco. 1987. Legume winter cover crops. Adv. Soil Sci. 7:95–139.

Scott, T.W., and R.F. Burt. 1985. Cover crops and intercrops of New York. p. 452.00. *In* Cornell Coop. Ext. Fact Sheet.

Touchton, J.T., W.A. Gardner, W.L. Hargrove, and R.R. Duncan. 1982. Reseeding crimson clover as a N source for no-tillage grain sorghum production. Agron. J. 74:283–287.

----, D.H. Rickerl, R.H. Walker, and C.E. Snipes. 1984. Winter legumes as a nitrogen source for no-tillage cotton. Soil Tillage Res. 4:391–401.

Triplett, G.B., Jr., D.M. Van Doren, Jr., and B.L. Schmidt. 1968. Effect of corn (*Zea mays* L.) stover mulch on no-tillage corn yield and water infiltration. Agron. J. 60:236–239.

Unger, P.W. 1978. Straw-mulch rate effect on soil water storage and sorghum yield. Soil Sci. Soc. Am. J. 42:486–491.

Utomo, M. 1986. Role of legume cover crops in no-tillage and conventional tillage corn production. Ph.D. diss., Univ. of Kentucky, Lexington.

Varco, J.J. 1986. Tillage effects on transformation of legume and fertilizer nitrogen and crop recovery of residue nitrogen. Ph.D. diss. Univ. of Kentucky, Lexington (Diss. Abstr. 8705310).

Wagger, M.G. 1987. Timing effects of cover crop desiccation on decomposition rates and subsequent nitrogen uptake by corn. p. 35–37. *In* J.F. Power (ed.) The role of legumes in conservation tillage systems. Soil Conserv. Soc. Am., Ankeny, IA.

Wischmeier, W.H., and J.V. Mannering. 1965. Effect of organic matter content of the soil on infiltration. J. Soil Water Conserv. 20:150–152.

Worsham, A.D., and R.H. White. 1987. Legume effects on weed control in conservation tillage. p. 113–119. *In* J.F. Power (ed.) The role of legumes in conservation tillage systems. Soil Conserv. Soc. Am., Ankeny, IA.

Wright, A.J., and E. Coxworth. 1987. Benefits from pulses in the cropping systems of northern Canada. p. 108 *In* J.F. Power (ed.) The role of legumes in conservation tillage systems. Soil Conserv. Soc. Am., Ankeny, IA.

9 Exploiting Forage Legumes for Nitrogen Contribution in Cropping Systems[1]

O. B. Hesterman

Michigan State University
East Lansing, Michigan

For many years, farmers have used forage legumes in cropping systems to take advantage of their N contribution to succeeding crops. It has long been recognized that including a forage legume in a cropping system can enhance the productivity of a succeeding nonlegume crop. It is also generally agreed that at least a portion of this productivity enhancement is due to N contribution from the legume (Stickler et al., 1958; Bolton et al., 1976; Baldock et al., 1981; Martin and Touchton, 1983; Bruulsema and Christie, 1987; Hesterman et al., 1987; Scott et al., 1987). In agronomic studies that examined the benefit of forage legumes in crop rotations, a 10 to 50% yield increase was observed for the nonlegumes following the legumes when compared to a nonlegume monoculture (Stickler et al., 1958; Bolton et al., 1976; Higgs et al., 1976; Baldock et al., 1981; Voss and Shrader, 1984; Hesterman et al., 1986; Bruulsema and Christie, 1987).

Some researchers have concluded that N contributions alone are responsible for these yield improvements (Sutherland et al., 1961; Baldock and Musgrave, 1980). However, many others believe that the benefits of forage legumes in cropping systems are due to some combination of N contribution and yield-enhancing effects not directly associated with N (rotation effects) (Baldock et al., 1981; Hesterman and Sheaffer, 1984; Russelle et al., 1987). In long-term experiments, corn (*Zea mays* L.) grown in rotation with other crops produced 5 to 10% greater yields than corn following corn when fertility management was optimal for both systems (Higgs et al., 1976; Voss and Shrader, 1979; Kurtz et al., 1984). As much as 25% of the corn yield increase in rotations has been attributed to rotation effects (Baldock et al., 1981). Some postulated causes of these rotation effects include improvement in soil structure (Rynasiewicz, 1945; Odland and Sheenan, 1957; Barber, 1972), reduction in disease (Curl, 1963) or phytotoxic substances (Barber, 1972), and production of growth-promoting substances (Ries et al., 1977).

[1] Contribution of Dep. of Crop and Soil Sciences, Michigan State Univ., East Lansing, MI 48824. Michigan Agric. Exp. Stn. Journal Article 12465.

Cropping Strategies for Efficient Use of Water and Nitrogen, Special Publication no. 51.

Although many researchers have documented the total yield-enhancing contribution of forage legumes in cropping systems, and most researchers agree that both N and rotation effects are involved, a major challenge continues to be quantifying these beneficial effects. Many researchers have reported experimental measurements of legume-N contributions (that portion of the total contribution due to N). The first section of this chapter will concentrate on the methods employed, the results obtained, and associated problems with the methods. The second section presents results from experiments in which various cropping strategies were tested for their effectiveness in exploiting legume-N contributions to subsequent nonlegume crops.

QUANTIFYING NITROGEN CONTRIBUTIONS

Three methods have been used by researchers over the years to measure the amount of N contributed from a forage legume to a nonlegume in a cropping system: (i) measurement of the N in the legume biomass; (ii) calculation of an N-fertilizer replacement value for the legume; and (iii) measurement of N contribution by tagging the legume with ^{15}N and following that N tag into the subsequent crop. For these three methods, legume-N contribution is comprised of *both* biologically fixed N and soil N that is merely being cycled through the legume plant. Few attempts have been made to differentiate between these two legume N sources.

Nitrogen in Legume Biomass

With the legume biomass method, the researcher measures the quantity of biomass incorporated into the soil (herbage + roots) and the N concentration of that biomass. A simple calculation results in a measure of the quantity of N (in kg ha^{-1}) produced by the legume, available for incorporation into the soil, and potentially usable by the following crop. Experiments using this method have resulted in N contribution values between 18 (Smith, 1956) and 390 (Groya and Sheaffer, 1985) kg of N ha^{-1} from a single season's growth of a forage legume. Variance in N contribution using this method can be attributed to, among other factors, legume species, age of the legume stand, and the harvest management employed during the year of measurement. Of the three methods of quantifying N contribution, this is perhaps the easiest. However, there is a significant limitation to using this measurement as an estimation of N contribution. Not all of the legume-N produced is mineralized and available to the subsequent crop. In fact, some studies suggest that only 10 to 30% of the N incorporated in legume material is absorbed by the following crop (Ladd et al., 1983; Ladd and Amato, 1986; Harris and Hesterman, 1987), with the remainder being accounted for in soil organic matter, in the inorganic soil N pool, and by losses from the system due to denitrification and leaching.

Nitrogen Fertilizer Replacement Value

The N-fertilizer replacement value is defined as the quantity of fertilizer N required to produce a yield in a crop that does not follow a legume that is identical to that produced by incorporation of the legume. The procedure to quantify the N-fertilizer replacement value involves: (i) generation of an N-fertilizer response curve for a nonlegume monoculture; and (ii) comparison of the yield of a nonlegume crop following a legume without N-fertilizer application with yield from the response curve. Figure 9–1 is a typical N-fertilizer response curve, and demonstrates the use of this method. In the example, corn following a legume produced a yield of 6.5 Mg ha^{-1}. The equivalent amount of N-fertilizer required to produce that 6.5 Mg ha^{-1} of corn without the legume is 75 kg of N ha^{-1}. Thus, the legume is credited for an N value of 75 kg ha^{-1}, and that is considered the N-fertilizer replacement value of the legume.

Many researchers have measured N-fertilizer replacement values of legumes and have reported N contribution values between 24 (Stickler et al., 1958) and 176 kg of N ha^{-1} (Boawn et al., 1963). Although the N-fertilizer replacement value may give a more accurate estimate of N contribution than estimating contribution based on biomass N production, this method assumes

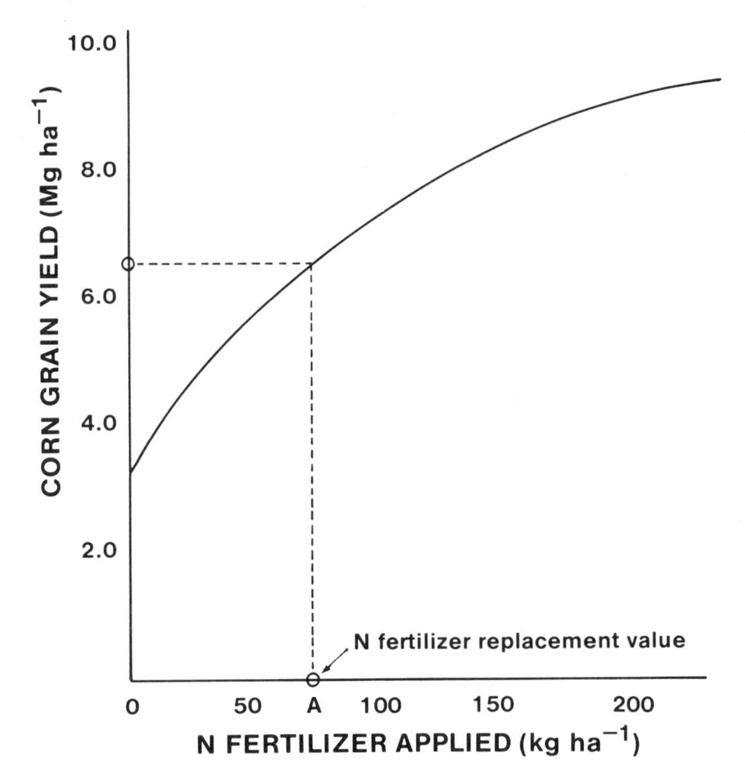

Fig. 9–1. Typical N-fertilizer response curve with N-fertilizer replacement value example (point A on x-axis) where maize grain yield following the legume = 6.5 Mg ha^{-1}.

that increased nonlegume yield is solely due to legume N contribution and that fertilizer N and residue N are equally available. The importance of beneficial effects other than N has been previously stated in this chapter.

Estimates of legume residue-N availability during the first subsequent cropping year have ranged from 10 to 34% (Fribourg and Bartholomew, 1956; Yaacob and Blair, 1980; Ladd et al., 1981 and 1983; Harris and Hesterman, 1987). These values are lower than typical fertilizer N-use efficiency values of 40 to 70% (Stanford, 1973). Inorganic fertilizer N appears to be more readily available than organic legume sources for absorption by the crop. This is likely due to the necessity of microbial turnover of legume N before it can be absorbed by plant roots.

Tag Legume with Nitrogen-15

With this method, the researcher fertilizes growing legume plants with the stable isotope ^{15}N, thus preventing nodulation and enriching the legume material with this heavy isotope. By incorporating the enriched legume plant material into the soil, the recovery of ^{15}N in a subsequent crop can be measured. This method results in a direct measurement of N contribution. Experiments using this method have shown that between 10 and 30% of legume N is absorbed by the subsequent crop in a crop rotation. Using this method, Ladd et al. (1983) found that the first wheat (*Triticum aestivum* L.) crop following soil amendment with ^{15}N-enriched *Medicago littoralis* L. residue recovered 27.8% of the legume N. In a subsequent experiment, Ladd and Amato (1986) found recoveries between 16 and 18%. In recent studies in Michigan, Harris and Hesterman (1987) found that 16 to 25% of alfalfa (*Medicago sativa* L.) residue N was recovered by a subsequent corn crop in field plots.

Although the ^{15}N method is a more accurate measure of N contribution than either of the previously described methods, several problems still exist. These include: (i) high cost of ^{15}N, resulting in limitation on plot size and scope of experiments; (ii) high cost of analyzing plant and soil samples for ^{15}N enrichment; and (iii) when nonlegume recovery of legume N is determined by measuring ^{15}N uptake from labeled residues, mineralization-immobilization turnover can result in low apparent recovery rates (Jansson and Persson, 1982).

In summary, researchers have quantified legume N contribution in cropping systems, especially using the N-fertilizer replacement value and N in legume-biomass methods. Future needs in the area of measuring N contributions include: (i) the need to distinguish between N contributions and rotation effects; and (ii) the need to refine ^{15}N methodologies so that an understanding can be gained of the rate and extent of mineralization-immobilization turnover.

CROPPING STRATEGIES TO EXPLOIT NITROGEN CONTRIBUTIONS

Many cropping strategies can be used to exploit the N contributions of forage legumes in cropping systems. In this section, five different strategies will be described along with results from selected experiments that typify the response that can be expected. In all cases, corn has been chosen as the nonlegume "subsequent" crop, while these experiments document results using different legume species. The N-fertilizer replacement value method was commonly used to estimate legume-N contributions in the following cropping strategy experiments. Therefore, when N-contribution values are presented, they will be in terms of N-fertilizer replacement values.

Conventional Rotations

In conventional crop rotations, the forage legume is grown for 2 or more years and is harvested each year as forage. After this extended period of forage production, the legume is then either killed by chemical treatment or incorporated into the soil using tillage equipment. The nonlegume is seeded into this area the following growing season.

Many studies have been conducted with conventional rotation systems and this is perhaps the most common farming practice used to exploit N contribution from forage legumes. Results from selected experiments show that, without N fertilizer, corn yields were increased by 2.5 to 6.4 Mg ha^{-1} when corn was planted following a forage legume vs. following corn. These values represent yield increases of 80 to 260% (Table 9–1). Fertilizer replacement values ranged from 84 to 110 kg of N ha^{-1}. In these same experiments, when N fertilizer was applied to corn following both corn and the legume, the average yield increase was only 18% (Table 9–2). One indication of the N-supplying capability of the legume is a comparison of corn yield (with N fertilizer applied) following corn vs. corn yield (without N fertilizer) following the legume. For the experiments listed in Tables 9–1 and 9–2, the average yield increase for the rotation (no N fertilizer on corn) compared to the monoculture (with N fertilizer) was only 3%.

Table 9–1. Corn grain yield (with no N fertilizer applied) following either corn or multiple years of a forage legume, percentage yield increase, and N-fertilizer replacement value (FRV) of the forage legumes from selected experiments.

Reference	Location	Corn yield following:		Yield increase	FRV
		Corn	Legume		
		—— Mg ha^{-1} ——		%	kg N ha^{-1}
Bolton et al., 1976	ONT	1.5	5.4	260	110
Baldock et al., 1981	WI	4.4	7.9	80	84
Sutherland et al., 1961	IA	3.0	5.5	83	89–108
Voss and Shrader, 1984	IA	3.5	9.9	183	

Table 9-2. Corn grain yield (with and without N-fertilizer applications) following either corn or multiple years of a forage legume, and percentage yield increase (decrease) from selected experiments.

Reference	Location	Nitrogen fertilizer applied to corn	Corn yield following:			Yield increase (decrease) for corn following legume	
				Legume			
			Corn	+N†	−N‡	+N†	−N‡
		kg ha^{-1}	Mg ha^{-1}			%	
Bolton et al., 1976	ONT	110§	5.3	6.9	5.4	30	2
Baldock et al., 1981	WI	168¶	7.4	8.5	7.9	15	7
Sutherland et al., 1961	IA	134¶	6.0	--	5.5	--	−8
Voss and Shrader, 1984	IA	200¶	9.0	9.8	9.9	9	10

† Corn following legume received supplemental N fertilizer at stated rate.
‡ Corn following legume received no supplemental N fertilizer.
§ Only fertilizer N rate used in experiment.
¶ Fertilizer-N rate associated with highest continuous corn yield.

Table 9-3. Corn grain yield (with no N fertilizer applied) following either corn or 1 yr of a green manure, percentage yield increase, and N-fertilizer replacement value (FRV) of the green manure from selected experiments.

Reference	Location	Corn yield following:		Yield increase	FRV
		Corn	Green manure		
		Mg ha^{-1}		%	kg N ha^{-1}
Stickler et al., 1958	IA	3.8	4.9	29	34
Bruulsema and Christie, 1987	ONT	6.5	8.6	32	90–125
Hesterman et al., 1986	MN	3.7	7.3	97	69–126
Voss and Shrader, 1979	IA	3.5	7.1	103	

Table 9-4. Corn grain yield (with and without N-fertilizer applications) following either corn or 1 yr of a green manure, and percentage yield increase (decrease) from selected experiments.

Reference	Location	Nitrogen fertilizer applied to corn	Corn yield following:			Yield increase (decrease) for corn following green manure	
				Green manure			
			Corn	+N†	−N‡	+N†	−N‡
		kg ha^{-1}	Mg ha^{-1}			%	
Stickler et al., 1958	IA	56§	6.0	--	4.9	--	−18
Bruulsema and Christie, 1987	ONT	224§	9.5	--	8.6	--	−9
Hesterman et al., 1986	MN	168§	6.5	8.6	7.3	32	12
Voss and Shrader, 1984	IA	200§	9.0	9.5	7.1	6	−21

† Corn following green manure received supplemental N fertilizer at stated rate.
‡ Corn following green manure received no supplemental N fertilizer.
§ Fertilizer-N rate associated with highest continuous corn yield.

Green Manure

In a green manure system, the legume is grown for 1 yr only and is not harvested. After that single year, the nonlegume is planted after either killing the legume with a chemical or, more commonly, incorporating the legume using tillage equipment. Results from selected experiments show that, without N fertilizer, corn yields were increased by 1.1 to 3.6 Mg ha^{-1} when corn was planted following a green manure vs. following corn (Table 9-3). These values represent yield increases of 32 to 103%. Fertilizer replacement values ranged from 34 to 126 kg of N ha^{-1}. In two of the selected experiments (Voss and Shrader, 1984; Hesterman et al., 1986), N fertilizer was applied to corn following both corn and the green manure, and the average yield increase for the green manure system was 19% (Table 9-4). Corn yield in the green manure system (no N fertilizer on corn) averaged 9% *less* than corn yield in the monoculture system (with N fertilizer) (Table 9-4).

Annual Alfalfa

When grown as an annual in the North Central states, alfalfa is seeded in early spring and harvested for forage at immature (bud) stages. Annual alfalfa refers to a system in which alfalfa is grown for a single year, incorporated into the soil, and, after that year, a nonlegume is planted. Annual alfalfa is not an annual species; it is the perennial *Medicago sativa* L., managed as an annual. Managing alfalfa in this manner in the Northern areas of the USA makes it possible to use nondormant alfalfa in annual crop rotations. Nondormant alfalfas, while providing greater fall growth and N_2-fixation potentials than the more adapted dormant alfalfas, will not likely survive the harsh winter conditions in the North.

In four studies in which corn was planted following annual alfalfa and corn grain yield was compared to corn yield following corn, when no N fertilizer was applied, corn yields were increased by 1.8 to 2.9 Mg ha^{-1} (Table 9-5). These values represent yield increases of 55 to 120%. In three of the four experiments, N fertilizer was applied as a treatment to corn following both corn and annual alfalfa, and the average yield increase for the annual

Table 9-5. Corn grain yield (no N fertilizer applied) following either corn or 1 yr of annual alfalfa, percentage yield increase, and N-fertilizer replacement value (FRV) of the annual alfalfa from selected experiments.

		Corn yield following:			
Reference	Location	Corn	Annual alfalfa	Yield increase	FRV
		—— Mg ha^{-1} ——		%	kg N ha^{-1}
Bolton et al., 1976	ONT	1.5	3.3	120	
Bruulsema and Christie, 1987	ONT	4.4	6.8	55	
Baldock et al., 1981	WI	4.4	7.3	66	51
Hesterman et al., 1986	MN	3.7	6.4	73	31-81

Table 9-6. Corn grain yield (with and without N-fertilizer applications) following either corn or 1 yr of annual alfalfa, and percentage yield increase (decrease) from selected experiments.

Reference	Location	Nitrogen fertilizer applied to corn	Corn yield following:			Yield increase (decrease) for corn following annual alfalfa	
				Annual alfalfa			
			Corn	+N†	−N‡	+N†	−N‡
		kg ha^{-1}	Mg ha^{-1}			%	
Bolton et al., 1976	ONT	110§	5.3	6.4	3.3	21	−38
Bruulsema and Christie, 1987	ONT	224¶	9.5	--	6.8	--	−28
Baldock et al., 1981	WI	168¶	7.4	7.9	7.3	7	−1
Hesterman et al., 1986	MN	168¶	6.5	8.4	6.4	29	−2

† Corn following annual alfalfa received supplemental N fertilizer at stated rate.
‡ Corn following annual alfalfa received no supplemental N fertilizer.
§ Only fertilizer-N rate used in experiment.
¶ Fertilizer-N rate associated with highest continuous corn yield.

alfalfa system was 19% (Table 9-6). Corn yield in the annual alfalfa system (no N fertilizer on corn) averaged 17% *less* than corn yield in the monoculture system (with N fertilizer) (Table 9-6).

Overseeding

In this system, a forage legume is seeded into a small grain crop [commonly winter wheat (*Triticum aestivum* L.) in Michigan] in early spring. It initially establishes underneath the expanding small grain canopy. After the small grain is harvested, the legume continues to grow for the remainder of that season, and is then either killed by tillage or a chemical treatment that fall or the following spring. The nonlegume (corn) is then planted into that area. In the overseeding system, the legume is grown in combination with the small grain. Only two studies were found in which this overseeding technique was tested (Table 9-7). In one study (Bruulsema and Christie, 1987) in Ontario, Canada, corn following 1 yr of a legume overseeded into oat (*Avena sativa* L.) outyielded corn following oat with no overseeded legume by 2 Mg

Table 9-7. Corn grain yield (no N fertilizer applied) following a small grain either with or without an overseeded legume, percentage yield increase, and fertilizer replacement value (FRV) of the overseeded legume from selected experiments.

Reference	Location	Corn yield following:		Yield increase	FRV
		Small grain/no legume	Small grain/with legume		
		Mg ha^{-1}		%	kg N ha^{-1}
Bruulsema and Christie, 1987	ONT	6.5	8.5	31	90–125
Williams and Hesterman, 1987, unpublished data	MI	4.3	6.2	44	70–140

Table 9-8. Corn grain yield (with and without N-fertilizer applications) following a small grain either with or without an overseeded legume, and percentage yield increase (decrease) from selected experiments.

Reference	Location	Nitrogen fertilizer applied to corn	Corn yield following: Small grain/no legume	Small grain/ with legume +N†	Small grain/ with legume −N‡	Yield increase (decrease) for corn following legume +N†	−N‡
		kg ha^{-1}	Mg ha^{-1}			%	
Bruulsema and Christie, 1987	ONT	224§	9.5	--	8.5	--	−11
Williams and Hesterman, 1987, unpublished data	MI	78§	6.1	6.6	6.2	8	2

† Corn following overseeded small grain received supplemental N fertilizer at stated rate.
‡ Corn following overseeded small grain received no supplemental N fertilizer.
§ Fertilizer-N rate associated with highest corn yield following small grain/no legume.

ha^{-1} (a 31% increase) (Table 9-7). Similar results were found, although yield levels were lower, in studies conducted for 2 yr in Michigan (Williams and Hesterman, 1987, unpublished). In both of these studies, the fertilizer replacement value of the overseeded legume was measured and the values ranged between 70 and 140 kg of N ha^{-1}.

When N fertilizer was applied as a treatment to corn following a small grain both with and without an overseeded legume, Williams and Hesterman found a yield increase of 8% in favor of the overseeding system. Corn yield in the overseeding system (no N fertilizer) was either 11% less (Bruulsema and Christie, 1987) or 2% greater (Williams and Hesterman, 1987, unpublished) than corn yield in the small grain-corn system with no overseeded legume (with N fertilizer) (Table 9-8).

Interseeding

In an interseeding (or interplanting) system, the forage legume is planted in combination with the nonlegume. Unlike a pasture system, the legume is grown primarily as an N source for the nonlegume. Three studies were found in which an interseeding system was tested with corn as the nonlegume crop (Tables 9-9 and 9-10). In two of the experiments, with no N fertilizer applied, corn interseeded with a legume outyielded corn grown alone by either

Table 9-9. Corn grain yield (no N fertilizer applied) either with or without an interseeded legume, percentage yield increase, and fertilizer replacement value (FRV) of the interseeded legume from selected experiments.

Reference	Location	System: Corn alone	System: Interseeded legume	Yield increase	FRV
		Mg ha^{-1}		%	kg N ha^{-1}
Triplett, 1962	OH	1.6	2.8	75	--
Scott et al., 1987	NY	2.9	3.2	10	17

Table 9-10. Corn grain yield (with and without N-fertilizer applications) either with or without an interseeded legume, and percentage yield increase (decrease) from selected experiments.

Reference	Location	Nitrogen fertilizer applied to corn	System			Yield increase (decrease) for interseeded system	
			Corn alone	Interseeded legume			
				+N†	−N‡	+N†	−N‡
		kg ha^{-1}	Mg ha^{-1}			%	
Triplett, 1962	OH	224§	6.1	5.8	2.8	−10	−54
Nordquist and Wicks, 1974	NB	134§	6.8	4.7	--	−31	--
Scott et al., 1987	NY	95¶	4.6	--	3.2	--	−30

† The interseeded corn system received supplemental N fertilizer at stated rate.
‡ The interseeded corn system received no supplemental N fertilizer.
§ Only fertilizer-N rate used in experiment.
¶ Fertilizer-N rate associated with highest "corn alone" yield.

10 (Scott et al., 1987) or 75% (Triplett, 1962) (Table 9-9). When N fertilizer was applied as a treatment to corn, both with and without an interseeded legume, Triplett (1962) reported a 10% yield decrease for the interseeded system and Nordquist and Wicks (1974) reported a 31% yield decrease for the interseeded system (Table 9-10). Corn yield in the interseeding system (no N fertilizer) was either 30% less (Scott et al., 1987) or 54% less (Triplett, 1962) than corn yield with no interseeded legume (with N fertilizer) (Table 9-10). Only one study reported the N-fertilizer replacement value of the interseeded legume, and it was found to be only 17 kg of N ha^{-1} (Table 9-9).

SUMMARY AND CONCLUSION

For the studies cited above, in situations in which no N fertilizer was applied, most cropping systems benefited from inclusion of a forage legume, with corn yield increases averaging between 38 and 152% for the different cropping strategies. When, however, N fertilizer was used, only the conventional rotations consistently demonstrated a positive corn yield response to the legume. Of the more "novel" systems discussed, overseeding may hold potential for exploiting the contribution of legume N, but more research with this system is needed. Unless N fertilizer is not available, interseeding systems do not show promise.

To conclude, farmers have been exploiting forage legumes for N contribution in cropping systems for many centuries. Scientists have recently been able to quantify these N contributions using various methods, some with more accuracy and success than others. There are some novel systems for exploiting forage legumes in cropping systems that have been tested recently also with varied success. The potential for further exploitation of forage legumes for N contributions is great, however, to further exploit these legumes both scientists and farmers need to know more about legume species other than those few that researchers have tested to this point. We need to know more about the interactions of legumes in the systems with different methods of tillage, with different pest problems, and with different pest-control strategies. Finally, and perhaps most importantly in terms of adopting these systems on a large scale, we need to know more about the economics of legume N compared with the economics of N from synthetic fertilizer sources under various fertilizer costs and commodity price situations. Only when it can be demonstrated that cropping systems that utilize legume N are more profitable will these systems become more commonplace.

REFERENCES

Baldock, J.O., R.L. Higgs, W.H. Paulson, J.A. Jackobs, and W.D. Shrader. 1981. Legume and mineral N effects on crop yields in several crop sequences in the upper Mississippi valley. Agron. J. 73:885–890.

----, and R.B. Musgrave. 1980. Manure and mineral fertilizer effects in continuous and rotational crop sequence in central New York. Agron. J. 72:511–518.

Barber, S.A. 1972. Relation of weather to the influence of hay crops on subsequent corn yields on a Chalmers silt loam. Agron. J. 64:8–10.

Boawn, L.C., J.L. Nelson, and C.L. Crawford. 1963. Residual nitrogen from NH_4NO_3 fertilizer and from alfalfa plowed under. Agron. J. 55:231–235.

Bolton, E.F., V.A. Dirks, and J.W. Aylesworth. 1976. Some effects of alfalfa, fertilizer and lime on corn yield in rotations on clay soil during a range of seasonal moisture conditions. Can. J. Soil Sci. 56:21–25.

Bruulsema, T.W., and R.B. Christie. 1987. Nitrogen contribution to succeeding corn from alfalfa and red clover. Agron. J. 79:96–100.

Curl, E.A. 1963. Control of plant diseases by crop rotation. Bot. Rev. 29:413–479.

Fribourg, H.A., and W.V. Bartholomew. 1956. Availability of nitrogen from crop residues during the first and second season after application. Soil Sci. Soc.. Am. Proc. 20:505–508.

Groya, F.L., and C.C. Sheaffer. 1985. Nitrogen from forage legumes: Harvest and tillage effects. Agron. J. 77:105–109.

Harris, G.H., and O.B. Hesterman. 1987. Recovery of nitrogen-15 from labeled alfalfa residue by a subsequent corn crop. p. 58–59. *In* The role of legumes in conservation tillage systems. Proc. Natl. Conf., Athens, GA. 27–29 April. Soil Conserv. Soc. Am., Ankeny, IA.

Hesterman, O.B., M.B. Russelle, C.C. Sheaffer, and G.H. Heichel. 1987. Nitrogen utilization from fertilizer and legume residues in legume-corn rotations. Agron. J. 79:726–731.

----, and C.C. Sheaffer. 1984. Evaluation of N contributions and rotation effects in legume-corn sequences. p. 128. *In* Agronomy abstract. ASA, Madison, WI.

----, ----, D.K. Barnes, W.E. Leuschen, and J.H. Ford. 1986. Alfalfa dry matter and nitrogen production, and fertilizer N response in legume-corn rotations. Agron. J. 78:10–23.

Higgs, R.L., W.H. Paulson, J.W. Pendleton, A.F. Peterson, J.A. Jackobs, and W.D. Shrader. 1976. Crop rotations and nitrogen: Crop sequence comparisons on soils of the driftless area of southwestern Wisconsin. 1967–1974. Univ. of Wisconsin College of Agric. and Life Sci. Res. Bull. R2761.

Jansson, S.L., and J. Persson. 1982. Mineralization and immobilization of soil nitrogen. *In* F.J. Stevenson (ed.) Nitrogen in agricultural soils. Agronomy 22:229–252.

Kurtz, L.T., L.V. Boone, T.R. Peck, and R.G. Hoeft. 1984. Crop rotations for efficient nitrogen use. p. 295–306. *In* R.D. Hauck (ed.) Nitrogen in crop production. ASA, CSSA, and SSSA, Madison, WI.

Ladd, J.N., and M. Amato. 1986. The fate of nitrogen from legume and fertilizer sources in soil successively cropped with wheat under field conditions. Soil Biol. Biochem. 18:417–425.

----, ----, R.B. Jackson, and J.H.A. Butler. 1983. Utilization by wheat crops of N from legume residues decomposing in soils in the field. Soil Biol. Biochem. 15:231.

----, J.M. Oades, and M. Amato. 1981. Distribution and recovery of nitrogen from legume residues decomposing in soils sown to wheat in the field. Soil Biol. Biochem. 13:251–256.

Martin, G.W., and J.T. Touchton. 1983. Legumes as a cover crop and source of nitrogen. J. Soil Water Conserv. (May–June):214–216.

Nordquist, D.T., and G.A. Wicks. 1974. Establishment methods for alfalfa in irrigated corn. Agron. J. 66:377–380.

Odland, J.E., and J.E. Sheenan. 1957. The effect of redtop and red clover on yield of following crops of potatoes. Am. Potato J. 34:282–284.

Ries, S.K., V. Wert, C.C. Sweeley, and R.A. Leavitt. 1977. Triacontanol: A new naturally occurring plant growth regulator. Science 195:1339–1341.

Russelle, M.P., O.B. Hesterman, C.C.Sheaffer, and G.H. Heichel. 1987. Estimating nitrogen and rotation effects in legume-corn rotations. p. 41–42. *In* The role of legumes in conservation tillage systems. Proc. Natl. Conf., Athens, GA. 27–29 April. Soil Conserv. Soc. Am., Ankeny, IA.

Rynasiewicz, J. 1945. Soil aggregation and onion yields. Soil Sci. 60:387–395.

Scott, T.W., J. Mt. Pleasant, R.F. Burt, and D.J. Otis. 1987. Contributions of ground cover, dry matter, and nitrogen from intercrops and cover crops in a corn polyculture system. Agron. J. 7:792–798.

Smith, D. 1956. Influence of fall cutting in the seeding year on the dry matter and nitrogen yields of legumes. Agron. J. 48:236–239.

Stanford, G. 1973. Rationale for optimum nitrogen fertilization in corn production. J. Environ. Qual. 2:159–166.

Stickler, F.C., W.D. Shrader, and I.J. Johnson. 1958. Comparative value of legume and fertilizer nitrogen in corn production. Agron. J. 50:157–160.

Sutherland, W.N., W.D. Shrader, and J.T. Pesek. 1961. Efficiency of legume residue nitrogen and inorganic nitrogen in corn production. Agron. J. 53:339–342.

Triplett, G.B., Jr. 1962. Intercrops in corn and soybean cropping systems. Agron. J. 54:106–109.

Voss, R.D., and W.D. Shrader. 1979. Crop rotations: Effect on yields and response to nitrogen. Iowa State Univ. Coop. Ext. Serv. Pm-905.

----, ----. 1984. Rotation effects and legume sources of nitrogen for corn. *In* D.F. Bezdicek et al. (ed.) Organic farming: Current technology and its role in a sustainable agriculture. Agronomy 46:61–68.

Yaacob, O., and G.J. Blair. 1980. The growth and nitrogen uptake of Rhodesgrass grown on soils with various histories of legume cropping. Plant Soil 57:249–255.

10 Role of Cropping Systems in Environmental Quality: Groundwater Nitrogen[1]

James S. Schepers
USDA-ARS
University of Nebraska-Lincoln
Lincoln, Nebraska

Nitrate (NO_3) contamination of groundwater is causing producers, policymakers, and researchers to reevaluate all phases of agricultural production from the simplest cultural practice to entire cropping systems. The threat of groundwater contamination by NO_3 has been evident for several decades in many parts of the USA. Hindsight indicates that society should have been more concerned when agricultural chemicals began showing up in groundwater, for example, contamination from potato (*Solanum tuberosum* L.) production on Long Island. History further indicates that all too often we have treated reports of groundwater contamination by agricultural sources as isolated problems that have little application to other areas. In reality, circumstances contributing to the problem probably should have been evaluated more thoroughly with the intent for application to other situations. In this regard, society seems to have followed a reactive strategy rather than taking a more proactive approach. The tendency to react to a crisis rather than to a perceived problem will probably not change because of economic considerations unless the threat of groundwater contamination becomes greater or society dictates a change in strategy and a willingness to ultimately bear the cost.

It is doubtful that society today is any more intelligent or ingenious than our predecessors, but we can take advantage of new technology and hopefully benefit from past experiences. To capitalize on previous experiences and effectively utilize new technology requires integration of more and more information. Computers have greatly expanded our ability to synthesize large amounts of data into predictive models, and the effective use of expert systems is on the horizon. If agriculture is to realistically deal with the challenge of groundwater contamination and make a significant impact on the problem, it will require something more than a "routine physical." The outcome of

[1] Contributed from the USDA-ARS in cooperation with the Agric. Res. Div., Univ. of Nebraska-Lincoln, Lincoln, NE 68583.

Cropping Strategies for Efficient Use of Water and Nitrogen, Special Publication no. 51.

such a comprehensive examination may be some rather drastic changes in agricultural production practices and economic considerations. In some cases, current agricultural production entities may even have to bear the burden of past cultural practices that have been shown to degrade groundwater quality.

Evaluating a cropping system for its influence on groundwater quality may involve several levels of detail depending on whether the intent is to simply determine if a given system represents an environmental threat or if the intent is to minimize the impact of an existing problem. In reality, all cropping systems represent some threat to the environment as do certain native conditions. Unfortunately, society frequently associates native conditions with a pristine environment. But as noted by Lewis and Clark, many rivers in the Midwest and Great Plains carried high sediment loads prior to settlement by pioneers. This example illustrates that an examination of cropping systems for their influence on groundwater quality should begin by dispensing all preconceived ideas because what may be good for one aspect of the environment may be harmful to another. The premise to be taken in this chapter is that N application rates must be carefully adjusted to the individual cropping system, climatic area, soil phase, and farmers' management capability to reduce groundwater contamination. Such sweeping statements must be tempered by further stating that N fertilizer is only one potential source of NO_3 and that some cropping systems may already be environmentally sound and have minimal impact on groundwater quality.

REASONS FOR NITRATE LEACHING

A reasonable starting point when evaluating the impact of cropping systems on groundwater quality is to examine the potential sources of contamination and the forces of nature driving the various processes. Sources of NO_3 found in soil include precipitation, waste by-products to include animal manure and septic tanks, irrigation water, organic matter decomposition, fertilizer, and natural deposits. Regardless of the N form applied, it can be converted to NO_3 in soil.

Simply stated, leaching of NO_3 to the groundwater occurs when the water content of soil exceeds the soil's ability to retain the water. Nitrate is known to leach through soils and into groundwater within a matter of days in some cases, while it may require many years for other ecosystems. Because of the dynamics of N and water, an annual budget of each may not be appropriate and therefore cropping systems of the future must be sensitive to all aspects of the hydrologic and N cycling on a short- and long-term basis.

Precipitation patterns, including irrigation practices, may have a major influence on NO_3 leaching. The obvious reason for irrigating is insufficient water for the crop, but in most cases irrigation is only intended to supplement rainfall except where irrigation is used to leach salts from the root zone. The relative influence of precipitation vs. irrigation on NO_3 leaching can be seen by comparing monthly precipitation and evapotranspiration (ET) for

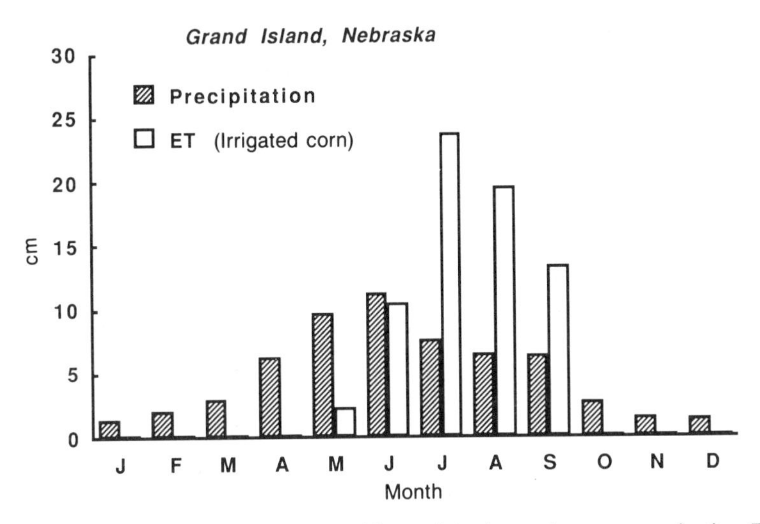

Fig. 10–1. Long-term (30-yr) average monthly precipitation and evapotranspiration (ET) for irrigated corn at Grand Island, NE (Ruffner and Bair, 1985).

a given crop (Fig. 10–1). Not only is the need for irrigation obvious, but the potential for leaching during the fall, winter, and early spring may be great. The example of excess fall through spring precipitation given in Fig. 10–1 may be common for many of the Northern states where temperatures do not permit year-round cropping (Table 10–1). Frozen soils retard leaching and promote runoff, but once soils thaw in the spring, NO_3 may be rapidly flushed from the root zone or denitrified. The period from May through September is usually considered the growing season for high N-use crops such as corn (*Zea mays* L.) and sorghum [*Sorghum bicolor* (L.) Moench]. October through April precipitation accounted for 35% (Aberdeen, SD) to 57% (Lexington, KY) of mean annual precipitation for selected locations in the Eastern, Midwest, and Great Plains states (Table 10–1). Precipitation amount and seasonal distribution do not necessarily characterize locations likely to have NO_3 leaching problems because climate and soil type have a major influence on leaching. Some locations where NO_3 contamination of groundwater would be expected may not create problems because excessive winter precipitation, lack of freezing temperatures, and coarse-textured soils permit adequate dilution of NO_3 in groundwater.

Many soils where corn and sorghum are grown are thawed by 1 April or before, however, the amount of precipitation is not offset by ET until mid-June. Considering this fact, 1 April through 15 June may be an opportune time for NO_3 leaching. At least 20% and up to 34% of mean annual precipiation for the locations in Table 10–1 occurred during this period. Precipitation amounts between 1 April and 15 June for most of these locations were in the range of 150 to 250 mm, which is adequate to initiate NO_3 leaching within the root zone and perhaps into the vadose zone. Simulated NO_3 leaching in a loamy sand soil from the Sandhills of Nebraska using recorded precipitation for the years 1973 to 1984 indicated that 20 to 80%

Table 10-1. Precipitation during selected months for several U.S. locations. After J.A. Ruffner and F.E. Blair (1985).

Location	Oct.–April	1 April–15 June	May–Sept.	Annual mean	Fractional distribution† Oct.–April	Fractional distribution† 1 April–15 June
	mm					
Aberdeen, SD	167	157	309	476	0.35	0.33
Dodge City, KS	176	166	343	519	0.34	0.32
St. Paul, MN	247	193	435	682	0.36	0.28
Grand Island, NE	210	210	416	625	0.34	0.34
Des Moines, IA	315	239	490	805	0.39	0.30
Madison, WI	325	209	454	779	0.42	0.27
Columbia, MO	444	272	518	962	0.46	0.28
Peoria, IL	424	241	468	892	0.48	0.27
Raleigh, NC	581	235	559	1140	0.51	0.21
Lansing, MI	389	201	390	779	0.50	0.26
Albany, NY	483	202	445	928	0.52	0.22
Allentown, PA	609	246	506	1115	0.55	0.22
Columbus, OH	510	222	433	943	0.54	0.24
Indianapolis, IN	550	246	463	1013	0.54	0.24
Lexington, KY	631	245	478	1109	0.57	0.22
Pendleton, OR	240	67	91	331	0.73	0.20
Boise, ID	227	72	81	308	0.74	0.23

† Fraction of precipitation for each time period is relative to the mean annual precipitation.

of the residual soil N in soil on 1 April would be leached below the root zone by 15 June (Schepers and Martin, 1987). The scenario was even worse for a fine sand-textured soil where 60 to 100% of the residual N would have been leached under these climatic conditions.

Nitrogen and water management problems are compounded on sandy soils with low water-holding capacity. One cannot assume that leaching only occurs when no crops are growing because irrigation may be excessive or the precipitation may fall at a time when it exceeds the available moisture-holding capacity of the soil. Likewise, the growing season may end with little or no root zone water depletion. As a result, storage capacity in the soil for winter precipitation may be reduced and the potential for leaching or runoff increased.

The potential for NO_3 leaching may be accelerated because of variations in monthly precipitation patterns, especially if no crop is growing. Both high-intensity rainfall and above-average amounts of precipitation increase the potential for N losses via runoff and leaching. Relatively large amounts of monthly precipitation will keep the soil water status above field capacity for a longer time and thereby increase NO_3 leaching as well as denitrification in poorly drained soils. This problem is significant in the Midwest and Great Plains where spring precipitation is quite variable just prior to planting corn and sorghum. For example, rainfall for Hastings, NE averages 110 mm during May, but in 1965 they received 317 mm with 99 mm coming in 1 d. Long-term records using a γ-distribution for this location indicate that there is a 50% probability of receiving 97 mm of rainfall during May, but there is also

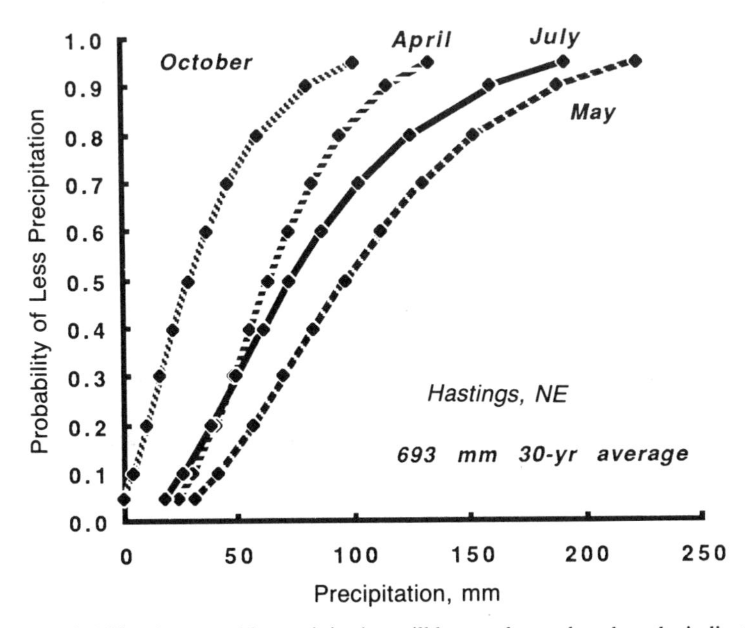

Fig. 10–2. Probability that monthly precipitation will be equal to or less than the indicated precipitation amount for Hastings, NE.

a 20% probability of receiving more than 153 mm (Fig. 10–2). Such uncertainty places a great deal of importance on fertilizer management decisions because N-fertilizer costs are typically lower in the fall of the year and time for N application is usually more abundant than prior to planting. Although many studies indicate the greatest fertilizer N recovery by corn from sidedress applications (Anderson et al., 1982; Russelle et al., 1983), the fear of wet weather and inability to make timely sidedress N applications causes many producers to strongly consider fall or spring applications. Even when preplant applications are used, fear of wet weather causes many producers to apply additional N to make up for possible leaching losses prior to crop uptake.

CULTURAL PRACTICES

It may appear that there are no solutions to this dilemma, but nitrification inhibitors can be used to delay transformation of anhydrous ammonia fertilizer to NO_3. Use of nitrification inhibitors may not be appropriate on all soils or with all anhydrous ammonia applications because of a lack of synchronization between soil N availability and crop N requirement. Other approaches include fertigation and high clearance fertilizer applicators. One such applicator is being developed at Iowa State University using spoke injectors mounted on a special high clearance tractor (Baker et al., 1985). An alternative approach to minimizing NO_3 leaching prior to planting is to create a soil environment that is more conducive to storage of fall and spring precipitation. Producers with irrigation can accomplish this by ending the growing season with as large a soil water deficit as possible. Secondly, producers could be encouraged to plant fall cover crops such as rye (*Secale cereale* L.), barley (*Hordeum vulgare* L.), or similar crops that grow rapidly and can survive low winter temperatures. Winter-annual cover crops would take up soil NO_3 that otherwise might be leached and they deplete the soil of water which thereby increases the water storage capacity of the soil. Leguminous cover crops such as vetch (*Vicia* spp.) may also be functional in terms of NO_3 recovery and soil water depletion. The N_2-fixing capability of legumes also makes them a desirable cover crop in climates where fall and winter temperatures do not severely limit growth. The economic incentive to use fall cover crops and especially the more expensive legumes may not be immediately obvious, but over time such practices have been shown to improve soil tilth and increase yields (Olson et al., 1986). Cover crops also serve a vital function by stabilizing the soil and reducing wind and water erosion.

Fallow during the growing season may increase the potential for NO_3 leaching in several ways. Nitrate leaching is usually minimized in regions where fallowing is necessary to accumulate enough soil water to support crop growth on alternate years. Fallowing to control weeds or growth of a volunteer crop can create a situation conducive to NO_3 leaching when no more water can be retained in the root zone. To further accentuate the problem, as soil water content increases to near field capacity, N mineralization in surface soil increases (Doran, 1982). By coincidence or otherwise, the optimum soil water content for mineralization nearly corresponds to the soil water status where leaching could be initiated by additional rainfall. Therefore, when the soil water content is near field capacity, NO_3 is in a vulnerable situation. Such a condition is frequently encountered on fallow land. An extensive example of this is the USDA-Agricultural Stabilization and Conservation Service payment-in-kind (PIK) program. Under the PIK program, many producers idled 20% or more of their corn acreage with the stipulation that residue cover must be provided and weeds must be controlled. Weed control involving tillage aerates the soil which in itself may increase mineralization and increase infiltration, but tillage may also dry the soil that can decrease mineralization, so the net influence of tillage on mineralization cannot be generalized. The fact remains however, that mineralization proceeds whether there is a growing crop or not, and that NO_3 produced is subject to leaching.

Cover crops on land that would otherwise be fallow provide erosion control and also tend to retard NO_2 leaching by incorporating the N into plant biomass. This approach has merit as long as soil moisture storage is not the primary reason for fallowing. One possible scenario to reduce NO_3 leaching from fallow land is to plant a cover crop such as oat (*Avena sativa* L.) in the spring, allow the crop to mature and reseed itself to provide lush growth for grazing by livestock in the fall. A large portion of the N contained in residue would be mineralized during the following growing season and would therefore function as a slow-release N source. Dolan and Schepers (1985) found that oat planted in early May on set-aside land in Nebraska produced 2.8 Mg/ha of dry matter by mid-September and accumulated 70 kg of N/ha that might have otherwise been leached under adverse climatic conditions. In this study, corn was planted adjacent to either oat or fallow to assess root exploration by corn into set-aside land. Residual soil N in the fall at a distance of one row (0.76 m) and two rows (1.52 m) from the corn indicated minimal N buildup and downward NO_3 movement near the corn compared to greater NO_3 accumulation and movement further from the corn (Fig. 10–3). Much of the N uptake by the oat crop apparently came from the surface 60 cm where mineralization would be expected to be the greatest. Although little NO_3 leaching was indicated below 120 cm between the spring and fall sampling, approximatey 170 kg/ha of N was mineralized during the growing season. While this value may appear large, it is less than that reported for the same site under irrigated corn production (Schepers and Mielke, 1983). The major challenge to researchers and producers is to develop and implement production practices that are culturally compatible and environmentally synchronized.

Mineralized N can be utilized by the crop during the growing season reducing the need for fertilizer N, provided the supply of N and crop requirements are synchronized. This synchronization requires an understanding of the processes regulating mineralization and the dynamics of crop N uptake. The amount and timing of N-fertilizer application can then be adjusted to supplement other N sources. While these considerations may seem fundamental to good N management, implementation of desired cultural practices may not be physically possible. As a result, cultural practices are usually less ideal than those designed to minimize NO_3 leaching. At other times, government policy may not encourage cultural practices that are known

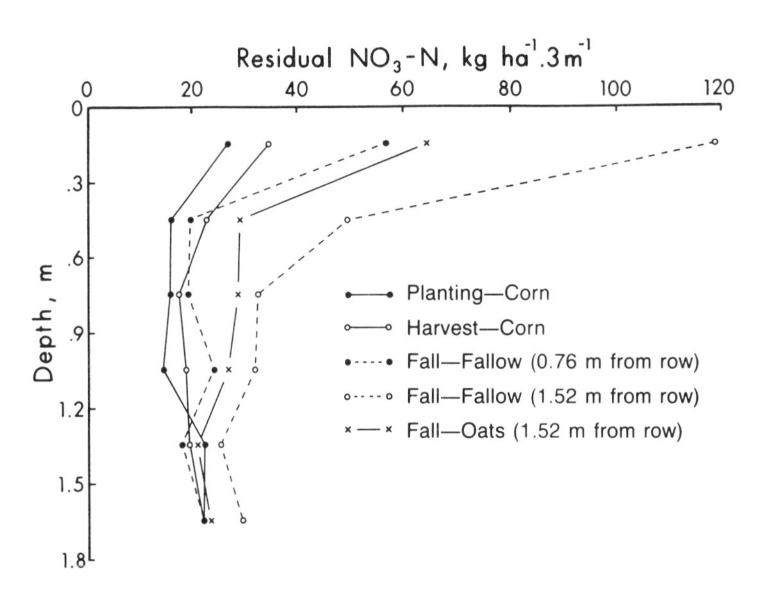

Fig. 10–3. Residual NO_3-N distribution in soil at planting (May) and at harvest (September) for irrigated corn, oat cover-crop, and fallow at two distances from corn border row, Grand Island, NE, 1983.

to reduce NO_2 leaching. A case in point is the PIK program where land must be diverted from corn production. Land fallowed because of the PIK program continued to mineralize N; however, guidelines were not provided or regulations implemented to utilize mineralized or residual N by cover crops.

Cover crops certainly reduce the potential for both surface water and groundwater contamination by NO_3, but strip cropping (fallow or cover crop between crop under production) can also be used to reduce potential NO_3 leaching. To be effective, strip cropping would require a modification in the minimal acceptable width of set-aside land to qualify for diverted land. Under present guidelines, the minimal set-aside width is about 20 m, but would need to be reduced to approximately 5 m or less to permit the roots of the crop to scavenge NO_3 from the fallow area. Dolan and Schepers (1985) found that the outside row of corn adjacent to fallow took up 51% (78 kg of N/ha) more N than the second row adjacent to fallow. A portion of the increased N uptake by the outside row was undoubtedly because of the 75% increase in yield compared to the second row from the fallow. Such yield responses in border rows are frequently attributed to increased light and nutrient availability. In this case, the yield response of the outside border row was larger than might be expected because the corn only received 75 kg/ha of fertilizer N. The increased N uptake by the outside border row was supported by soil samples collected to a depth of 180 cm under fallow at a distance of 76 and 152 cm from the outside row, which contained 160 and 288 kg/ha of residual N, respectively (Fig. 10–3 and 10–4). These results indicate that if NO_3 leaching under fallow is of primary concern, then reducing the minimal acceptable width of fallow to 2 m would have the greatest influence on residual N and probably on NO_3 leaching. However, corn yields would increase con-

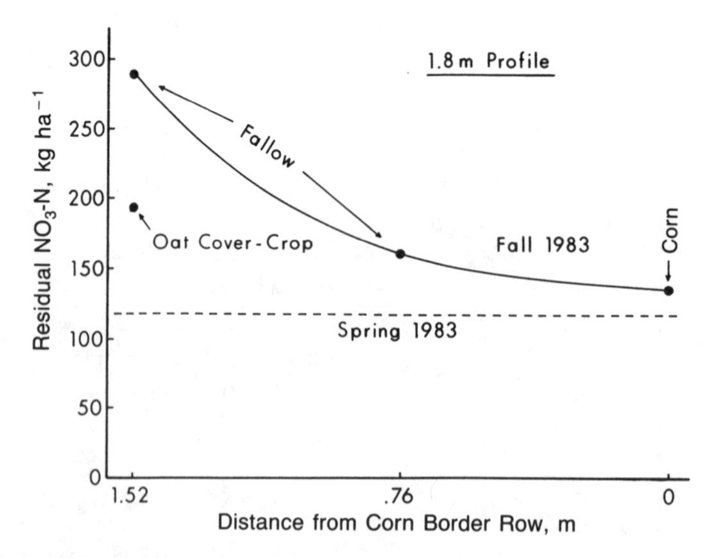

Fig. 10–4. Residual NO_3-N in 1.8-m depth at planting (May) and at harvest (September) for irrigated corn, and fallow at two distances from corn border row, Grand Island, NE, 1983.

siderably unless the narrow set-aside width is coupled with a reduction in-crop area.

If strip cropping involving such narrow widths were to become accepted, yields would increase because of the border effect. The yield increase could be offset by requiring a greater proportion of set-aside land. The net effect would be the same level of production with lower fertilizer costs because the outside row would require little, if any, fertilizer and there would be less potential leaching from the fallowed area. Strip cropping may be difficult for producers with large equipment, but perhaps not as difficult as it may seem for row crop producers who could disconnect some of the plant units. Producers using furrow irrigation could probably modify their water-management program to accommodate strip cropping, but those using center-pivot irrigation systems would find strip cropping economically infeasible and environmentally unsound.

Field borders may only represent one aspect of a cropping system, but they can reduce potential NO_3 contamination of streams and ultimately the groundwater. Lateral subsurface flow and perched water tables are common in areas where soils are shallow over consolidated rock or where other materials retard downward water movement. In the past, field borders and grassed waterways have been used to filter sediment from runoff and thereby reduce the sediment load in streams and rivers. Recently, thought also has been given to using field borders or riparian zones near spring-fed streams as possible scrubbers of the subsurface flow. Deep-rooted trees and grasses could intercept and conceivably remove some NO_3 in laterally moving water, provided the water was within the root zone and the vegetation remained actively growing most of the time.

Deep-rooted crops such as alfalfa can be used to reduce high NO_3 levels that may have accumulated after several years of annual cropping or in situations where crop failure has resulted in substantial quantities of residual soil N (Viets and Hageman, 1971). Alfalfa can also be used to reduce concentrations of residual N in soils under abandoned feedlots (Schuman and Elliott, 1978). It should be noted however, that large quantities of N may be mineralized when crops such as alfalfa or other legumes are killed. Therefore, crops following legumes will usually require less N fertilizer (Elliott et al., 1987; Lohry et al., 1987). Mineralization of legume residues usually occurs more rapidly than with cereal residues, so unless crops can utilize the mineralized N, it may be susceptible to leaching. While much of the N, biologically fixed by legumes such as soybean, is removed in the grain, N-rich residues readily decompose prior to the next growing season and seldom pose any problems with immobilization (Heichel, 1987). Timing of N applications relative to crop requirements can be important because of N immobilization when large quantities of alfalfa, clover, or similar biomass are killed near planting time. Similar immobilization problems can also result in temporary N deficiencies when cover crops are destroyed and incorporated immediately prior to planting. Problems occasionally encountered with immobilization can usually be overcome by anticipating crop N requirements and synchronizing the cultural practices accordingly.

An alternative approach to minimize the impact of immobilization on crop N availability is to "dribble" a band of concentrated N on the soil surface to minimize contact with surface residues or to inject N fertilizer below the microbially active surface soil. These N-placement techniques are frequently used with reduced-tillage systems where immobilization necessitates higher rates of surface-applied N for yields comparable with conventional-tillage systems. Even when fertilizer N application rates are similar between reduced- and conventional-tillage systems, the reduced-tillage system may pose an increased threat to groundwater NO_3 contamination. Reduced runoff and sediment losses frequently observed when residues remain on the soil surface coincide with increased infiltration and potential NO_3 leaching (Gilliam and Hoyt, 1987). Trade-offs between infiltration and runoff will depend on site-specific climate, soil and crop considerations.

In extreme cases, it may be necessary to protect groundwater used for municipal sources of drinking water by restricting application of N fertilizer. It is not known how large a control area would be required around municipal well fields, but such an approach could reduce NO_3 leaching. Examples of large-scale diversion of productive cropland to grass are infrequent, however one such case exists in the Platte River Valley near Grand Island NE, where fertile land was taken out of production in the early 1940s for construction of an ammunition depot. A large portion of this land was planted to grass while land around the depot was developed for irrigated corn production. By the mid-1970s, most irrigation wells around the depot were pumping water containing 15 mg of N/L as NO_3 or greater. Water pumped from within the restricted depot area typically contained <5 mg of N/L as NO_3. Conversion of cultivated land to grassland for purposes of groundwater protection should be carefully examined because groundwater hydrology and soils may be such that this approach would not give the desired results in the time frame allotted.

Economic considerations at the present time encourage N fertilization for near maximum yields of irrigated corn. Producers under rain-fed conditions are motivated by similar economic forces, but usually have a greater degree of uncertainty because of climatic variability. Because of climatic variability, producers are forced to hedge their production practices in an attempt to maximize profits and as such, fertilizer-N application rates probably err toward the excessive side. The net effects of overfertilization are record yields and an ever-increasing list of groundwater contamination problems. Economics aside, if producers would be willing to reduce their yield expectations to perhaps 95% of the maximum, and then fertilize accordingly, N-fertilizer rates would be substantially reduced. A 9-yr N rate study of irrigated corn production in Nebraska indicates fertilizer N rates could be reduced as much as one-third (to 100 kg of N/ha) if yield goals were reduced 5% below the maximum yield (Kenneth Frank, 1986, personal communication). Examples such as this indicate that the net effect of slightly lower yields on groundwater contamination by NO_3 could be positive. Little has yet been stated or proven regarding producer liability in cases of groundwater contamination, but with time it will certainly become more of an issue. If lower rates of N fertilization are accepted and become common, producers and the public must be aware that greater variations in yield will be likely a. that prices may fluctuate accordingly. The reason is that as a primary nutrient, ample or excessive N tends to mask problems or symptoms that might otherwise be seen as a nutrient deficiency. Therefore, as N application rates decline, both direct and indirect nutrient stresses on crops may result.

CONCLUSION

Nitrogen and water management practices to reduce NO_3 leaching and contamination of groundwater are sure to evolve with time, but in many parts of the USA the environment cannot wait for the more traditional adoption of improved production systems. It is not surprising that society is becoming frustrated with ever-increasing reports of NO_3 contamination of groundwater and what may appear as a lack of remedial measures on the part of producers. There is little doubt that agriculture must assume a major portion of responsibility for the problem and will therefore be expected to make a concerted effort to remedy the situation. This huge task will require a coordinated effort by all phases of agriculture and an increased awareness by the public. The key to accomplishing this task will be an effective system to disseminate information. Major educational responsibilities may be centered at the land grant universities, but support and endorsement for such programs must lie within the framework of the fertilizer industry. In spite of past and present efforts to make producers aware of the potential for NO_3 contamination of groundwater, more regulations may be required to overcome some of the attitudes related to crop production and the economic considerations involved in decision making.

REFERENCES

Anderson, C.K., L.R. Stone, and L.S. Murphy. 1982. Corn yield as influenced by in-season application of nitrogen with limited irrigation. Agron. J. 74:396–401.

Baker, J.L., T.S. Colvin, S.J. Marley, and M. Dawlebeit. 1985. Improved fertilizer management with a point-injection applicator. ASAE Paper 85-1516. ASAE, St. Joseph, MI.

Dolan, M.S., and J.S. Schepers. 1985. The effects of field border conditions on a simulated corn-fallow system. p. 5. *In* Agronomy abstracts. ASA, Madison, WI.

Doran, J.W. 1982. Tillage changes soil. Crops Soils 34:10–12.

Elliott, L.F., R.I. Papendick, and D.F. Bezdicek. 1987. Cropping practices using legumes with conservation tillage and soil benefits. p. 81–89. *In* J.F. Power (ed.) Role of legumes in conservation tillage systems. Soil Consev. Soc. Am., Ankeny, IA.

Gilliam, J.W., and G.D. Hoyt. 1987. Effect of conservation tillage on fate and transport of nitrogen. *In* T.J. Logan et al. (ed.) Effects of conservation tillage on groundwater quality. Lewis Publ., Chelsea, MI.

Heichel, G.H. 1987. Legumes as a source of nitrogen in conservation tillage systems. p. 29–34. *In* J.F. Power (ed.) Role of legumes in conservation tillage systems. Soil Conserv. Soc. Am., Ankeny, IA.

Lohry, R.D., M.D. Clegg, G.E. Varvel, and J.S. Schepers. 1987. Fertilizer use efficiency in a sorghum-soybean rotation. p. 38–39. *In* J.F. Power (ed.) Role of legumes in conservation tillage systems. Soil Conserv. Soc. Am., Ankeny, IA.

Olson, R.A., W.R. Raun, Y.S. Chun, and J. Skopp. 1986. Nitrogen management and interseeding effects on irrigated corn and sorghum and on soil strength. Agron. J. 78:856–862.

Ruffner, J.A., and F.E. Bair. 1985. Weather in U.S. cities. Gale Res. Co., Book Tower, Detroit, MI.

Russelle, M.P., R.D. Hauck, and R.A. Olson. 1983. Nitrogen accumulation rates in irrigated maize. Agron. J. 75:593–598.

Schepers, J.S., and D.L. Martin. 1987. Public perception of ground water quality and the producers dilemma. p. 399–411. *In* Proc. Agric. Impacts on Ground Water—A conference. Natl. Water Well Assoc., ASA, Soil Conserv. Soc. Am., Omaha, NE. 11–13 Aug. 1986. Well Journal Publ. Co., Dublin, OH.

----, and L.N. Mielke. 1983. Nitrogen fertilization, mineralization, and leaching under irrigation in the Midwest. p. 325–334. *In* R. Lowrance et al. (ed.) Nutrient cycling in agricultural ecosystems. Spec. Pub. 23. Univ. of Georgia, Athens.

Schuman, G.E., and L.F. Elliott. 1978. Cropping an abandoned feedlot to prevent deep percolation of nitrate-nitrogen. Soil Sci. 126:237–243.

Viets, F.G., Jr., and R.H. Hageman. 1971. Factors affecting the accumulation of nitrate in soil, water, and plants. Agric. Handb. 413. USDA-ARS, Washington, DC.

11 Role of Cropping Systems in Environmental Quality: Saline Seep Control[1]

A. D. Halvorson
USDA-ARS
Akron, Colorado

Saline seeps are occurring more frequently in dryland farming areas throughout the Great Plains (Ballantyne, 1963; Berg et al., 1986; Brown et al., 1987; Colburn, 1983; Doering and Sandoval, 1976b; Halvorson and Black, 1974; Neffendorf, 1978; Vander Pluym, 1978). They are caused by a combination of geologic, climatic, and cultural conditions. The term *saline seep* describes a salinization process accelerated by dryland-farming practices that utilize water inefficiently, allowing water to move through salt-laden substrata below the root zone. Saline seep is accepted to mean intermittent or continuous saline water discharge, at or near the soil-surface downslope from recharge areas under dryland conditions, which reduces or eliminates crop growth in the discharge area because of increased soluble-salt concentrations in the root zone. Saline seeps can be differentiated from other saline soil conditions by their recent and local origin, saturated root-zone profile, shallow water table, and sensitivity to precipitation and cropping systems (Brown et al., 1983).

The characteristics, hydrology, and causes of most saline seeps are similar regardless of geographic location (Berg et al., 1986; Brown et al., 1983; Doering and Sandoval, 1976a, b; Halvorson and Black, 1974; Vander Pluym, 1978). Native or naturally occurring vegetation has been removed and replaced with agricultural crops that have a lower water requirement. Precipitation received in excess of the soil-root zone storage capacity, primarily during fallow or noncrop periods, is the source of water. The crop-fallow system of dryland farming has contributed significantly to the development of the saline-seep problem in the Northern Great Plains, but is not the only cause (Brown et al., 1983; Christie et al., 1985; Halvorson and Black, 1974). Periods of high precipitation, restricted surface and subsurface drainage due to roads and/or pipeline construction, snow accumulations resulting in large drifts (i.e., windbreaks, roadways, etc.), gravelly and sandy soils, drainage ways, leaky ponds and dugouts, and crop failures can contribute to seep develop-

[1]Contribution from USDA-ARS, Akron, CO 80720.

Cropping Strategies for Efficient Use of Water and Nitrogen, Special Publication no. 51.

ment. The above factors combined with the right geologic conditions can result in the development of a saline seep after many years of cropping. Generalized diagrams of geologic conditions resulting in seep development in the Northern Great Plains are shown in Fig. 11-1. Seeps generally develop on sidehills or toe-slopes of rolling to undulating topography common to the Great Plains, where permeable geologic material is underlain by less-

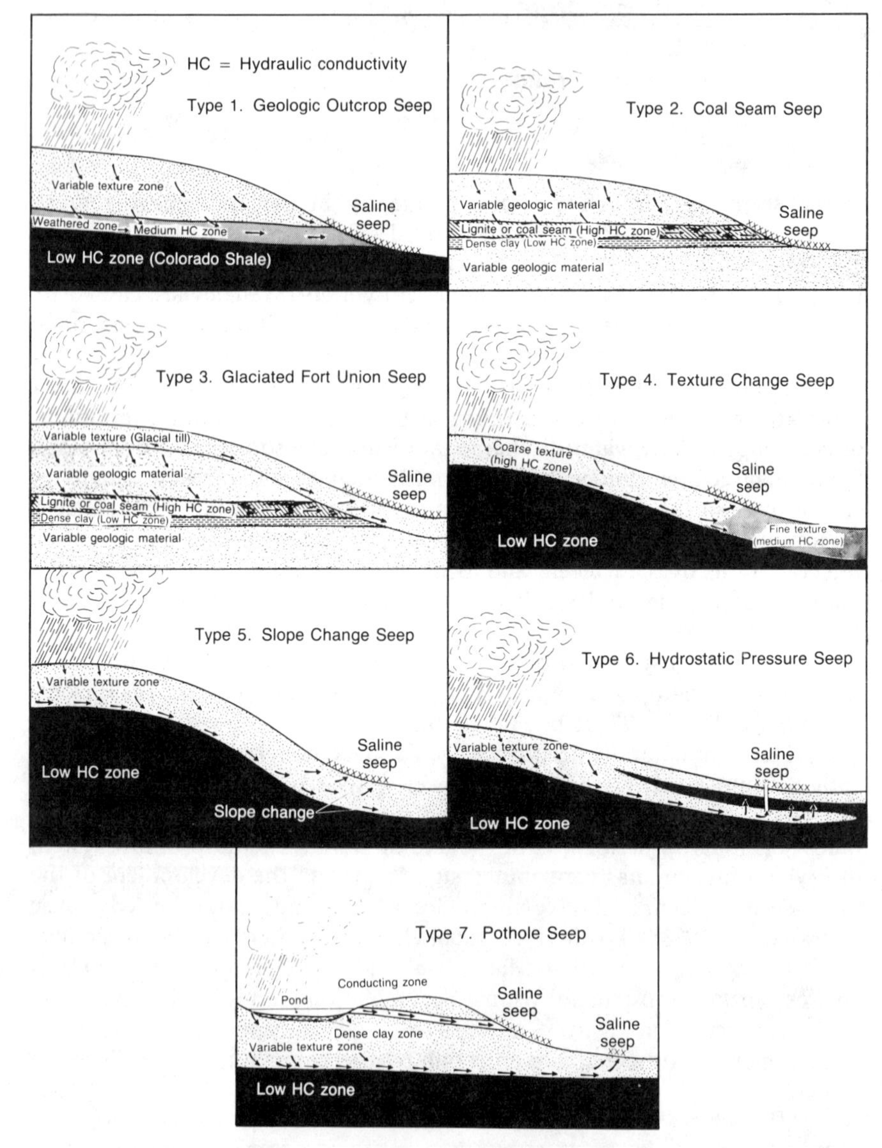

Fig. 11-1. Schematic diagrams illustrating seven geologic conditions for saline seep development (Brown et al., 1983).

permeable geologic strata. These geologic and topographic conditions result in conditions favorable to development of perched water tables.

GROUNDWATER QUALITY ASSOCIATED WITH SALINE SEEP

Hydrologic studies show that seeps are sustained by local recharge areas (Doering and Sandoval, 1976a; Halvorson and Black, 1974; Halvorson and Reule, 1980; Hendry and Schwartz, 1982; Naney et al., 1986). As the water-passes through the soil profile toward the perched or permanent water table, salts are dissolved and moved downward. Ferguson and Batteridge (1982) estimated that up to 90 Mg/ha of salt has been moved in the soil toward (to) the groundwater table in some areas of Montana. Christie et al. (1985) also reported a decrease in soil-profile salinity of cultivated land vs. that of an adjacent native noncultivated area, indicating movement of salt to lower depths. Doering and Sandoval (1981) reported the removal of 6.1 Mg/ha of salt and 50 kg of NO_3-N/ha from a drained seep area. These studies document the movement of soluble salts and NO_3-N toward and into shallow water tables.

Table 11-1 shows the composition of waters associated with several saline seeps in the Northern and Southern Great Plains. Note the high concentrations of NO_3 ($>$0.7 mmol/L) present in many of the shallow water tables associated with saline seeps. Much of the shallow groundwater associated with saline seeps is unsuitable for human and/or livestock consumption because of high NO_3 and salt levels, and for irrigation because of total salt concentration. Calcium, Mg, and Na sulfates are the dominate cations and anion in most of the shallow groundwater associated with saline seeps in the Great Plains. Chlorides are low compared to sulfates in the water and soil

Table 11-1. Chemical composition of waters associated with saline seeps in the Great Plains.

Location	pH	EC	Ca	Mg	Na	HCO_3	NO_3	Cl	SO_4	Source
		dS m^{-1}	mmol L^{-1}							
MT recharge	8.4	5	7	11	18	3.8	4.3	0.7	21	Halvorson and Black, 1974
MT seep	8.2	9	8	21	66	9.8	0.4	0.8	52	Halvorson and Black, 1974
MT seep	7.9	14	10	37	109	8.1	29.5	2.6	80	Halvorson and Black, 1974
MT seep	8.4	26	1	108	211	4.0	5.4	7.6	225	Miller, 1971
MT recharge	8.2	7	3	21	39	2.4	6.2	11.2	44	Miller, 1971
ND seep	3.7	10	9	36	59	--	5.7	2.1	70	Doering and Sandoval, 1981
ND seep	4.6	8	9	30	40	--	4.7	2.5	55	Doering and Sandoval, 1981
OK seep	8.1	5	15	16	26	--	0.6	12.3	27	Berg et al., 1986
OK seep	8.2	3	3	17	13	--	--	16.0	15	Naney et al., 1986

system of the Northern Great Plains. Chloride tends to be slightly higher in the Southern Great Plains. Soil chemistry studies show that soils in seep areas are generally in equilibrium with gypsum, lime, and other Ca-Mg sulfate-type minerals (Brun and Deutch, 1979; Doering and Sandoval, 1981; Oster and Halvorson, 1978; Timpson et al., 1986).

Studies in the Northern Great Plains have surmised that the NO_3 in the groundwater came primarily from two sources: (i) exchangeable NH_4 of geologic origin located deep in the profile oxidized to NO_3, and (ii) NO_3 leached from the root zone during fallow periods, which originated from mineralization of organic matter near the soil surface (Doering and Sandoval, 1981; Hendry et al., 1984; Power et al., 1974). Little, if any, of the NO_3 had its origin as fertilizer N because little fertilizer N had been used by dryland farmers in the Northern Great Plains prior to the early 1970s when saline seeps were becoming a prominent dryland salinity problem.

CONTROL METHODS

Since seeps are caused by water moving below the root zone in the recharge area, there will be no permanent solution to the saline-seep problem unless control measures are applied to the recharge area. There are two general procedures for managing seeps: (i) mechanically drain-ponded surface water where possible before it infiltrates, and/or intercept lateral flow of subsurface water with drains before it reaches the discharge area, and (ii) agronomically use the water before it percolates below the root zone. Each of these will be discussed in some detail with major emphasis on agronomic control.

Drainage

Undulating, near level land with poor surface drainage (potholes) can be recharge areas for saline seeps (Brown et al., 1983). Following heavy rain and rapid snowmelt, these potholes may fill with water temporarily. Where possible, surface drains should be installed to prevent the temporary ponding of surface water. Drainageways under roadbeds should be kept clear of debris and sediment so that they do not serve as sources of temporarily ponded surface water that infiltrates and contributes water to a saline seep. In the Central Great Plains, level bench terraces serve as temporary water impoundments that may be contributing water to saline seeps (Berg et al., 1986; Naney et al., 1986). Use of such water conservation practices may need to be evaluated if saline seep is a problem.

Drainage studies have shown that hydraulic control can be quickly accomplished with subsurface interceptor drains located on the upslope side of the seep area (Doering and Sandoval, 1976a; Sommerfeldt et al., 1978). However, a suitable outlet for disposal of the saline water needs to be available. Outlet considerations must include not only easement for transport of drainage water across intervening lands, but also the effect of drainage waters on the quality of streams or reservoirs that they might subsequently enter. The water is salt contaminated, usually high in NO_3, and disposal without downstream surface or groundwater pollution is difficult because of physical and legal constraints. Therefore, although effective, subsurface drainage is generally not satisfactory because of disposal problems and costs in the dryland crop areas of the Great Plains. Doering and Sandoval (1981) concluded that the best approach is to use the soil water for crop growth while it is a relatively nonsaline resource in the root zone of the recharge area.

Oosterveld (1978) used seep discharge water to irrigate the recharge area, thus recycling the salts back to the land. In general, the concept was successful, but limited water supplies for irrigation, cost of an irrigation system to deliver the water, and buildup of soil salinity in recharge area may reduce the practical application of this technique for saline seep control.

Agronomic Practices

Hydraulic control of saline seep areas can be effected agronomically by planting crops and utilizing cropping systems that will use available soil water in the root zone where it is a relatively nonsaline resource. This requires that recharge areas be distinctly identified and that farmers be willing and able to adopt the cultural practices needed for maximum soil water use and minimum percolation. Techniques for identifying recharge and potential saline seep areas have been developed (Alberta Agriculture, 1986; Brown et al., 1983; Halvorson and Rhoades, 1974; Halvorson and Rhoades, 1976). Early detection and diagnosis of a saline seep problem are important in designing and implementing effective crop management practices to prevent further damage or salinization. Any delay in implementing control practices can lead to a much larger and more difficult-to-manage problem. By early detection, a farmer may be able to change cropping systems to minimize the damage. Techniques for identifying potential seep and recharge areas include: (i) visual observations, (ii) field measurement of soil electrical conductivity, (iii) soil survey and geologic information, and (iv) auger or core-drilling procedures (Brown et al., 1983).

Alfalfa (*Medicago sativa* L.), seeded in recharge areas, is an effective crop for gaining hydraulic control of seep discharge areas (Brown et al., 1983; Brun and Worcester, 1975; Halvorson and Reule, 1980). Halvorson and Reule (1980) found that alfalfa extracted more water from the soil profile than native grass sod or small grain crops (Fig. 11–2). Land that had been fallowed was at or near field-capacity water content, creating a potential for any additional precipitation to percolate below the root zone. Alfalfa depleted the soil water more than the other crops, creating a larger effective reservoir to store annual precipitation by using more water during the growing season, thus reducing the potential for water loss to deep percolation. This resulted in the saline seep or discharge area drying sufficiently to once again grow normal grain and forage crops (Halvorson, 1984). Similar results were observed by the author in Colorado, when a farmer near Akron established alfalfa in a seep recharge area in 1984 to bring an active saline seep area under

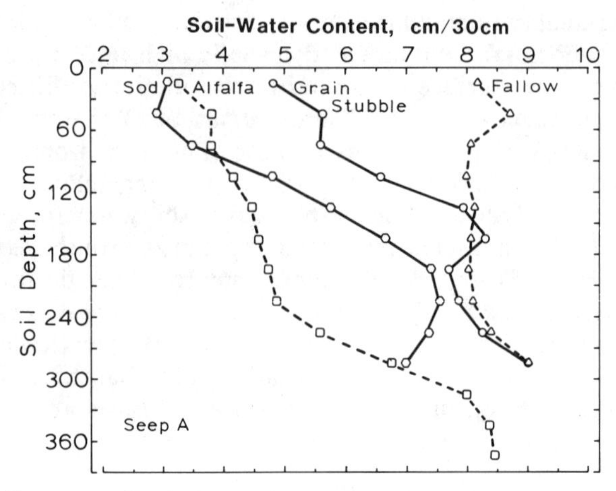

Fig. 11-2. Soil water profiles in September 1976 in the recharge area under native range (sod), alfalfa (seeded in 1973), spring wheat stubble, and fallow (Halvorson and Reule, 1980).

control. By the fall of 1985, the seep had dried sufficiently to allow the seep area to once again be crossed with farm machinery. In 1987, three cuttings of alfalfa were harvested from the discharge area where only salt-tolerant weeds grew in 1984.

Brown and Miller (1978) and Miller et al. (1981) also showed that alfalfa was effective for controlling saline seeps, while Brown (1983) further showed that it took 7 to 8 yr to recharge the dried soil profile to field-capacity water content when a fallow-winter wheat (*Triticum aestivum* L.)-barley (*Hordeum vulgare* L.) rotation followed 3 yr of alfalfa (Table 11-2). Similarly, Halvorson and Reule (1980) reported a rise in water table level where a farmer reverted back to a crop-fallow system of farming in the recharge area following several years of alfalfa production, during which hydraulic control of the seep area was achieved and the seep area was supporting near-normal crop production. These studies indicate that once a saline seep area has been

Table 11-2. Total soil water content (0-4.6 m) at the end of each growing season following 3 yr of alfalfa (Brown, 1983).

Year	Crop/fallow	Fall soil water	Annual precipitation
		mm 4.6 m^{-1}	mm
1973	Alfalfa, 3rd yr	217	--
1974	Fallow, no crop	342	278
1975	Winter wheat	330	563
1976	Barley	393	371
1977	Fallow, no crop	448	363
1978	Winter wheat	461	418
1979	Barley	461	208
1980	Fallow, no crop	524	380
1980	Estimated field capacity	573	--

controlled, reclaimed, and returned to normal crop production, a farmer cannot return to a conventional crop-fallow system of farming in the recharge area on a permanent basis. Soil water in the recharge area will need to be continually managed to prevent the recurrence of the saline seep.

Other work has shown that use of annual cropping systems with small-grain crops has resulted in control of seep areas (Alberta Agriculture, 1986; Bramlette, 1971; Halvorson and Reule, 1976; Holm, 1983; Steppuhn and Jenson, 1984). Use of annual, small-grain cropping systems to gain hydraulic control of seep discharge areas is a slower process than using alfalfa because of less soil water use and shallower rooting depths. Inclusion of oil seed crops such as safflower (*Carthamus tinctorius* L.) or sunflower (*Helianthus annuus* L.), which are normally deeper rooted than small grains (Table 11-3), will help deplete the stored soil water to greater depths, thereby increasing the capacity of the soil to store precipitation between crops or during fallow periods.

Black et al. (1981) describe several dryland cropping strategies for efficient water use to control saline seeps in the Northern Great Plains. They suggest using intensive, flexible cropping systems with adapted crops in combination with proper soil, water, and crop management practices to improve crop production-water-use relationships sufficiently to reduce the frequency or eliminate the need for summer fallow.

Flexible cropping involves planting a crop in years when stored soil water and expected growing season precipitation are sufficient to produce an economic crop yield. Summer fallowing is employed only when soil water and expected growing-season precipitation are not sufficient to produce a reasonable or economic crop yield. A farmer can assess the soil water supply by having a knowledge of soil texture and using a soil moisture probe to physically determine moist soil depth (Brown, 1958) or by any other method

Table 11-3. Rooting depth and soil water use by 11 dryland crops (Black et al., 1981).

	Fort Benton, MT		Culbertson, MT	
Crop	Rooting depth	Soil water use	Rooting depth	Soil water use
	m	mm	m	mm
Alfalfa, 1st yr	2.1	178	--	--
Alfalfa, 4th yr	5.5	666	--	--
Sanfoin, 1st yr	1.5	150	--	--
Sanfoin, 4th yr	4.0	561	--	--
Russian wildrye, 1st yr	2.1	318	--	--
Russian wildrye, 4th yr	3.0	475	--	--
Sweetclover, 1st yr	1.8	276	--	--
Sweetclover, 2nd yr	2.7	403	--	--
Safflower	2.2	249	2.1	229
Sunflower	2.0	206	--	--
Winter wheat	1.8	200	1.6	190
Rapeseed	1.5	170	--	--
Spring wheat	--	--	1.2	152
Barley	1.4	190	1.1	135
Dryland corn	1.2	94	--	--

of determining soil water. Based on available soil water and expected growing-season precipitation, a decision to crop or fallow can be made (Alberta Agriculture, 1986; Brown et al., 1981). Recropping or annual cropping is not recommended whenever plant-available soil water stored at planting time is less than about 76 mm (Alberta Agriculture, 1986; Black and Ford, 1976). Therefore, determining the soil water status of the root zone just before planting is essential if more intensive crop-management systems are to be successful. Halvorson and Kresge (1982) developed a computer model (FLEX-CROP) to help farmers decide the best cropping and soil-management options for wheat, barley (*H. sativum* Jess.), oat (*Avena sativa* L.), and safflower (*Carthamus tinctorius* L.) based on stored soil water and expected growing-season precipitation (program available from author). Weed control and soil fertility are also critical management factors in developing successful, flexible, dryland-cropping systems.

Black et al. (1974) and Halvorson and Black (1974) reported that 72 to 83% of the total water stored during the 84-week fallow period in the Northern Great Plains was stored during the first 36 weeks of the fallow period. Therefore, saving an additional 20 to 30 mm of water during the first overwinter period may eliminate the need for a summer-fallow period. Snow-management studies indicate that saving this much additional water during the first over-winter period is possible (Black et al., 1981; Black and Siddoway, 1976; Nicholaichuk and Gray, 1986).

Black and Siddoway (1976) found that precipitation use efficiency was 80% for a continuous cropping system vs. 50% for a spring wheat-winter wheat-fallow and 30% for a spring wheat-fallow rotation. Thus, with more intensive cropping, the potential for water loss to deep percolation in the Northern Great Plains is reduced.

Black et al. (1981) reported that crops grown under annual cropping systems used an average of 75 to 81% of the precipitation received between crop harvests within a grass barrier system; whereas, conventional spring wheat-fallow used only 40% (Table 11-4). The average amount of unused plant-available water between crops, a portion of which may contribute to saline seep development, averaged 473 mm for spring wheat-fallow and only 72 to 98 mm for annual cropping. These data demonstrate the potential of moving water, NO_3, and dissolved salts below the root zone in a spring wheat-fallow system compared with an annual cropping system. Adequate fertility is essential for optimizing yields in annual cropping systems (Black et al., 1982; de Jong and Halstead, 1986; Halvorson et al., 1976; Schneider et al., 1980).

If more intensive, flexible-cropping systems are to be successful, more efficient methods of storing soil water during noncrop periods is needed. In the Northern and Central Great Plains, soil water supplies can be increased by controlling weed growth and volunteer grain after harvest, leaving standing stubble to trap snow, utilizing annual or perennial barriers or windbreaks for snow trapping, and utilizing reduced and no-tillage systems (Black and Siddoway, 1976; Nicholaichuk and Gray, 1986; Smika and Whitfield, 1966). All of these practices will enhance the efficiency of soil water storage.

Table 11-4. Average precipitation-use efficiency per cropping sequence as influenced by cropping system within a tall wheatgrass barrier system over a 12-yr period (Black et al., 1981).

Cropping system	No. crops yr^{-1}	Total PPT $crop^{-1}$	Total water use $crop^{-1}$†	PUE‡	Annual grain yield No N	Annual grain yield +N	WUE No N	WUE +N
		——mm——		%	—kg ha^{-1}—		kg ha^{-1} mm^{-1}	
Annual cropping								
1. 6WW,B,S,B,WW,S,B	1.00	396	322	81	1328	1794	3.4	4.5
2. 5SW,S,B,WW,B,WW, B,WW	1.00	394	296	75	993	1822	2.5	4.6
3. 4SW,S,B,WW,S,SW, B,WW,B	1.00	390	318	82	969	1590	2.5	4.1
Three-yr rotation								
1. SW-WW-F	0.66	569	333	59	997	1416	2.6	3.7
Crop-fallow								
1. WW-F	0.50	788	404	51	1019	1247	2.6	3.1
2. SW-F	0.50	786	313	40	853	1065	2.2	2.7

† Crop water-use per crop is based on soil water use plus precipitation received from seeding to harvest.

‡ Symbols: +N = 34 kg of N ha^{-1} each crop year; PPT = precipitation; PUE = [(total water use/crop)/(total precipitation received/crop)] × 100; WUE = [(grain yield/ha)/(total precipitation/crop rotation)] × 100; WW = winter wheat; SW = spring wheat; B = spring barley; S = safflower; F = fallow.

However, if these practices are used, a more intensive cropping system than the conventional crop-fallow system must be employed or saline seep development will intensify.

CROPPING STRATEGIES FOR SALINE SEEP CONTROL

Crops differ in the amount of water required to produce an economical yield because of rooting depths and extraction patterns. Black et al. (1981) reported that safflower used more soil water to greater soil depths in 1 yr than any other annual dryland crop grown in Montana (Table 11-3). Alfalfa used only slightly less water the 1st yr than safflower or sweetclover (*Melilotus officinalis* L.), but alfalfa's ability to use growing-season precipitation plus existing-soil water supplies from progressively deeper soil depths in successive years marks alfalfa as the best crop for initial use to gain hydraulic control in recharge areas (Table 11-4). Sanfoin (*Onobrychio viriaefolia* Scop.) and Russian wildrye (*Elymus junceus* Fisch.) depleted soil water to a depth of 4 and 3 m, respectively, after 3 yr. Biennial sweetclover used more soil water than safflower the 1st yr and about equal to alfalfa the 2nd yr. Following alfalfa, sweetclover, and safflower, the ranking of crop water use in order of decreasing rooting depth and soil water use was sunflower (*Helianthus annuus* L.), winter wheat, rapeseed (*Brassica napus* L.), spring wheat, barley, and corn (*Zea mays* L.). Fallow becomes a viable option in flexible cropping systems once the soil water has been depleted from the root zone.

Selecting the best crop sequence requires knowledge of the amount and depth of soil water depletion by the previous crop. Crops should be grown in sequential order with increasing rooting depths until the depth and amount of soil water removal exceeds soil water recharge during noncrop periods (Black et al., 1982). An example would be the use of a spring wheat-winter wheat-fallow rotation, where the deeper rooted winter wheat follows spring wheat. A fallow period then follows winter wheat because of greater soil water depletion by this crop. Ideally, fallow should only be used when needed, such as following alfalfa, or when there is <76 mm of plant-available soil water at planting in the upper part of the root zone.

Successful recropping will require that crops be rotated in sequence to avoid specific weed, disease, and insect problems. Inclusion of oilseed crops in rotation with small grain allows the opportunity to use grass herbicides, helping control build-up of grassy weeds in the small grain crops (Bergman et al., 1979).

Soil fertility follows water in importance in a successful annual cropping system (Black et al., 1981; Black and Siddoway, 1976; de Jong and Halstead, 1986; Halvorson et al., 1976; Halvorson and Kresge, 1982; Schneider et al., 1980). As cropping frequency increases, the need for additional N increases proportionately (Fig. 11-3) and responses to P fertilizer become dependent upon soil-test P level and first satisfying crop N needs (Halvorson and Black, 1985). Nitrogen needs should be carefully balanced with expected plant-available water supplies and crop-yield potential.

The practice of a strict crop-fallow rotation restricts farmers to a fixed cropping system with limited flexibility to adjust cropping patterns to fit available water supplies. Selection of alternate cropping strategies to effectively use available water supplies requires a knowledge of the amount of water available at any given time, specific water requirements and rooting

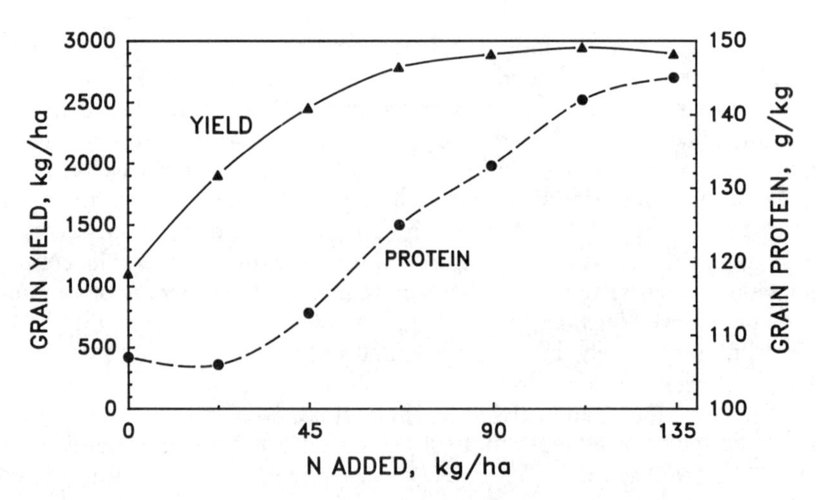

Fig. 11-3. Average recrop grain yield and protein content of winter wheat as affected by N-fertilization rate (Halvorson et al., 1976).

depths of adapted crops, and expected growing-season precipitation. A knowledge of the depth to some restricting or impermeable geologic strata and water table is essential if an effective cropping strategy to control or prevent the development of saline seeps is to be developed. Berg et al. (1986) and Naney et al. (1986) suggest that this type of strategy will also work in the Central and Southern Great Plains area to control saline seeps.

CONCLUSION

Based on the information presented in this chapter, the importance of cropping systems in controlling the dryland saline seep problem in the Great Plains has been shown. However, if more intensive cropping systems than the conventional crop-fallow system are to be used successfully, farmers need to know the soil water status at planting, an estimate of growing-season precipitation, crop-rooting depths and soil water-use characteristics, soil fertility, weed and pest control practices, water conservation strategies (i.e., snow management), yield and disease characteristics of crop varieties, crop rotations, and other soil and crop-management factors. Timely farm operations will be essential along with soil-test information and a knowledge of the depth to impermeable geologic strata and water tables to make a flexible, intensive crop-management system work. Cropping restrictions under the current federal farm program complicate the picture and make developing effective and economical cropping strategies difficult.

By developing and employing cropping systems that use water more efficiently, environmental quality will be preserved or improved in areas where dryland saline seeps exist as well as in areas where their potential for development occurs. Limiting the percolation of water below the root zone of dryland crops will reduce the movement of soluble salts and nitrates toward and into the groundwater, thus preserving its quality.

REFERENCES

Alberta Agriculture. 1986. Dryland saline seep control. Alberta Agric. AGDEX 518-11, Edmonton, Alberta, Canada.

Ballantyne, A.K. 1963. Recent accumulation of salts in the soils of southeastern Saskatchewan. Can. J. Soil Sci. 43:52–58.

Berg, W.A., C.R. Cail, D.M. Hungerford, J.W. Naney, and G.A. Sample. 1986. Saline seep on wheatland in northwest Oklahoma. p. 265–271. *In* Proc. Natl. Conf. on Ground Water Quality and Agricultural Practices. Lewis Publ. Co., Chelsea, MI.

Bergman, J.W., G.P. Hartman, A.L. Black, P.L. Brown, and N.R. Riveland. 1979. Safflower production guidelines. Montana Agric. Exp. Stn., Capsule Info. Ser., 8 (revised).

Black, A.L., P.L. Brown, A.D. Halvorson, and F.H. Siddoway. 1981. Dryland cropping strategies for efficient water-use to control saline seeps in the northern Great Plains, U.S.A. Agric. Water Manage. 4:295–311.

----, and R.H. Ford. 1976. Available water and soil fertility relationships for annual cropping systems. p. 286–290. *In* Proceedings regional saline seep control symposium. Montana State Univ., Bozeman, Coop. Ext. Serv. Bull. 1132.

----, and F.H. Siddoway. 1976. Dryland cropping sequences within a tall wheatgrass barrier system. J. Soil Water Conserv. 31:101–105.

----, ----, and J.K. Aase. 1982. Soil moisture use and crop management-(DRYLAND). p. 215–231. *In* Proc. Soil Salinity Conf., Lethbridge, Alberta, Canada. 29 November–2 December.

----, ----, and P.L. Brown. 1974. Summer fallow in the Northern Great Plains (winter wheat). p. 36–50. *In* Summer fallow in the western United States. USDA Conserv. Res. Rep. 17. U.S. Gov. Print. Office, Washington, DC.

Bramlette, G. 1971. Control of saline seeps by continuous cropping. *In* Proc. Saline Seep-Fallow Workshop, Great Falls, MT. 22–23 February. Highwood Alkali Control Assoc., Highwood, MT.

Brown, P.L. 1958. Soil moisture probe. U.S. Patent 2 860 515.

----. 1983. Saline seep control—soil water recharge under three rotations—following alfalfa. *In* Montana Chapter Soil Conserv. Soc. of America Meet., Bozeman, MT. 4–5 February.

----, A.L. Black, C.M. Smith, J.W. Enz, and J.M. Caprio. 1981. Soil water guidelines and precipitation probabilities for growing barley, spring wheat and winter wheat in flexible cropping systems in Montana and North Dakota. Montana Coop. Ext. Serv. Bull. 356.

----, H. Ferguson, and J. Holzer. 1987. Saline seep development and control in Montana. p. 28–33. *In* J.W. Bauder (ed.) A century of action: Natural resource development and conservation in Montana. Montana Chapter of Soil Conserv. Soc. Am., Bozeman.

----, A.D. Halvorson, F.H. Siddoway, H.F. Mayland, and M.R. Miller. 1983. Saline-seep diagnosis, control and reclamation. USDA Conserv. Res. Rep. 30.

----, and M.R. Miller. 1978. Soil and crop management practices to control saline seeps in the U.S. Northern Plains. p. 7.9–7.15. *In* Proc. of Meet. of Subcommission of Salt-Affected Soils, 11th Int. Soil Sci. Soc. Congr., Edmonton, Alberta, Canada. 21–24 June.

Brun, L.J., and R.L. Deutch. 1979. Chemical composition of salts associated with saline seeps in Stark and Hettinger Counties, North Dakota. N.D. Farm Res. 37(1):3–6.

----, and B.K. Worcester. 1975. Soil water extraction by alfalfa. Agron. J. 67:586–589.

Christie, H.W., D.N. Graveland, and C.J. Palmer. 1985. Soil and subsoil moisture accumulation due to dryland agriculture in southern Alberta. Can. J. Soil Sci. 65:805–810.

Colburn, E. 1983. Salt buildup in soil, slicing Texas yields. Crops Soils Magazine 35(4):26.

de Jong, E., and E.H. Halstead. 1986. Field crop management and innovative acres. p. 185–200. *In* Proc. Moisture Managment in Crop Production Conf., 18–20 November, Calgary, Alberta. Alberta Agriculture, Edmonton, Alberta.

Doering, E.J., and F.M. Sandoval. 1976a. Hydrology of saline seeps in the northern Great Plains. Trans. ASAE 19:856–861, 865.

----, and ----. 1976b. Saline-seep development on upland sites in the northern Great Plains. USDA ARS-NC-32.

----, and ----. 1981. Chemistry of seep drainage in southwestern North Dakota. Soil Sci. 132:142–149.

Ferguson, H., and T. Batteridge. 1982. Salt status of glacial till soils of north-central Montana as affected by the crop-fallow system of dryland farming. Soil Sci. Soc. Am. J. 46:807–810.

Halvorson, A.D. 1984. Saline-seep reclamation in the northern Great Plains. Trans. ASAE 27:773–778.

----, and A.L. Black. 1974. Saline-seep development in dryland soils of northeastern Montana. J. Soil Water Conserv. 29:77–81.

----, and ----. 1985. Long-term dryland crop responses to residual phosphorus fertilizer. Soil Sci. Soc. Am. J. 49:928–933.

----, ----, F. Sobolik, and N. Riveland. 1976. Proper management: Key to successful winter wheat recropping in Northern Great Plains. N.D. Farm Res. 33(4):3–9.

----, and P.O. Kresge. 1982. FLEXCROP: A dryland cropping systems model. USDA Production Res. Rep. 180.

----, and C.A. Reule. 1976. Controlling saline seeps by intensive cropping of recharge areas. p. 115–124. *In* Proceedings regional saline seep control symposium. Montana State Univ., Coop. Ext. Serv. Bull. 1132.

----, and ----. 1980. Alfalfa for hydrologic control of saline seep. Soil Sci. Soc. Am. J. 44:370–373.

----, and J.D. Rhoades. 1974. Assessing soil salinity in identifying potential saline-seep areas with field soil resistance measurements. Soil Sci. Soc. Am. Proc. 38:576–581.

----, and ----. 1976. Field mapping soil conductivity to delineate dryland saline seeps with four electrode technique. Soil Sci. Soc. Am. J. 40:571–575.

Hendry, M.J., R.G.L. McCready, and W.D. Gould. 1984. Distribution, source and evolution of nitrate in a glacial till of southern Alberta, Canada. J. Hydrol. 70:177–198.

----, and F. Schwartz. 1982. Hydrogeology of saline seeps. p. 25–40. *In* Proc. Soil Salinity Conf. Lethbridge, Alberta, Canada. 29 November–2 December.

Holm, H.M. 1983. Soil salinity: A study in crop tolerances and cropping practices. Saskatchewan Agric., Plant Industry Branch, Regina, Saskatchewan.

Miller, M.R. 1971. Hydrology of saline-seep spots in dryland farm areas—A preliminary evaluation. *In* Proc. Saline Seep-Fallow Workshop, Great Falls, MT. 22–23 February. Highwood Alkali Control Assoc., Highwood, MT.

----, P.L. Brown, J.J. Donovan, R.N. Bergatino, J.L. Sonderegger, and F.A. Schmidt. 1981. Saline seep development and control in the North American Great Plains-Hydrologic aspects. Agric. Water Manage. 4:115–141.

Naney, J.W., W.A. Berg, S.J. Smith, and G.A. Sample. 1986. Assessment of ground water quality in saline seeps. p. 274–285. *In* Proc. Agric. Impacts on Groundwater—A Conf. Omaha, NE. 11–13 August. Natl. Water Well Assoc., Dublin, OH.

Neffendorf, D.W. 1978. Statewide saline seep survey of Texas. M.S. thesis. Texas A&M Univ., College Station.

Nicholaichuk, W., and D.M. Gray. 1986. Snow trapping and moisture infiltration enhancement. p. 73–84. *In* Proc. Moisture Management in Crop Production Conf., Calgary, Alberta. 18–20 November, Alberta Agriculture, Edmonton, Alberta, Canada.

Oosterveld, M. 1978. Disposal of saline drain water by crop irrigation. p. 4.24–4.29. *In* Proc. Meet. of Subcommission of Salt-Affected Soils, 11 Int. Soil Sci. Soc. Congr., Edmonton, Alberta, Canada. 21–24 June.

Oster, J.D., and A.D. Halvorson. 1978. Saline seep chemistry. p. 2.7–2.29. *In* Proc. Meet. of Subcommission of Salt-Affected Soils, 11th Int. Soil Sci. Soc. Congr., Edmonton, Alberta, Canada. 21–24 June.

Power, J.F., J.J. Bond, F.M. Sandoval, and W.O. Willis. 1974. Nitrification in paleocene shale. Science 183:1077–1079.

Schneider, R.P., B.E. Johnson, and F. Sobolik. 1980. Saline seep management: Is continuous cropping an alternative? N.D. Farm Res. 37(5):29–31.

Smika, D.E., and C.J. Whitfield. 1966. Effect of standing wheat stubble on storage of winter precipitation. J. Soil Water Conserv. 21:138–141.

Sommerfeldt, T.G., H. Vander Pluym, and H. Christie. 1978. Drainage of dryland saline seeps in Alberta. p. 4.15–4.23. *In* Proc. Meet. of Subcommission of Salt-Affected Soils, 11th Int. Soil Sci. Soc. Congr., Edmonton, Alberta, Canada. 21–24 June.

Steppuhn, H., and D. Jenson. 1984. Barley can help control dryland salinity. Crops Soils Magazine 36(8):22–23.

Timpson, M.E., J.L. Richardson, L.P. Keller, and G.J. McCarthy. 1986. Evaporite mineralogy associated with saline seeps in southwestern North Dakota. Soil Sci. Soc. Am. J. 50:490–493.

Vander Pluym, H.S.A. 1978. Extent, causes and control of dryland saline seepage in the Northern Great Plains of North America. p. 1.48–1.58. *In* Proc. Meet. of Subcommission on Salt-Affected Soils, 11th Int. Soil Sci. Soc. Congr., Edmonton, Alberta, Canada, 21–24 June.

12 Low-Input Cropping Systems and Efficiency of Water and Nitrogen Use

J. K. Radke
USDA-ARS
Rodale Research Center
Kutztown, Pennsylvania

R. W. Andrews, R. R. Janke, and S. E. Peters
Rodale Research Center
Kutztown, Pennsylvania

In recent years, there has been increasing concern with modern, intensive farming methods, which rely heavily on pesticides and chemical fertilizers. While crop productivity has increased greatly in the last 30 yr, prevailing farming methods have been blamed for groundwater contamination with nitrates (NO_3) and pesticides (Hallberg, 1986), as well as surface water pollution (Myers, 1985). The safety of pesticides on the farm as well as pesticide residues on farm products are becoming a major issue as well (Pimentel et al., 1980; Tangley, 1986). The recent economic farm crisis has also raised interest in alternative methods, which can reduce input costs for the farmer, improving net returns (Lockeretz and Wernick, 1983).

In light of these issues, low-input agricultural systems are being investigated as alternatives to conventional systems. Farmers as well as consumers are asking that research efforts into these alternatives be increased (Tangley, 1986; Buttel et al., 1986; USDA, 1980).

Alternatives to conventional agricultural methods range over a wide spectrum. The USDA (1980) defines organic farming as. . .

> . . .a production system which avoids or largely excludes the use of synthetically compounded fertilizers, pesticides, growth regulators, and livestock feed additives. To the maximum extent feasible, organic farming systems rely upon crop rotations, crop residues, animal manures, legumes, green manures, off-farm organic wastes, mechanical cultivation, mineral-bearing rocks, and aspects of biological pest control to maintain soil productivity and tilth, to supply plant nutrients, and to control insects, weeds, and other pests.

Cropping Strategies for Efficient Use of Water and Nitrogen, Special Publication no. 51.

The term *low-input* will be used throughout this chapter. Our definition of low-input farming includes the USDA definition of organic farming, with the additional emphasis on using, to the greatest extent possible, resources generated on-farm, rather than resources produced externally.

With the increased interest in low-input farming, it becomes necessary to critically evaluate these systems. The purpose of this chapter is to discuss the implications of low-input systems for N and water-use efficiency.

Nitrogen-use efficiency (NUE) can be considered in terms of various criteria. An overview of several ways of defining of NUE is presented by Bock (1984). This chapter will deal primarily with crop yield efficiency, which refers to the relationship between yield and N application rate, and N recovery efficiency, which refers to the relationship between N recovered by the crop and N rate applied.

The definitions used by Bock refer to efficiency within 1 yr or one crop. In addition to NUE as measured for an individual plant or crop within a growing season, we believe that it is important to consider NUE over a period of several years, especially if crop rotation is an integral part of the farming system. There is a growing body of evidence for biological efficiencies associated with an increase in the active fraction of soil organic matter resulting from legume or animal manure use in cropping systems.

In this chapter, we focus on several management strategies that are common, although not necessarily universal, to low input systems. The greater part of this chapter is concerned with N-use efficiency. Water-use efficiency will be considered when discussing the results of the Rodale Research Center experiments.

MANAGEMENT STRATEGIES IN LOW-INPUT SYSTEMS

Minimal or No Pesticide Use

The absence of or minimal use of pesticides in low-input systems will be discussed only briefly in this chapter. While this is an important characteristic of low-input systems from an environmental and an economic standpoint, its relationship to NUE is less direct. If crop yield were severely reduced in a low-input system as a result of weed or insect problems, NUE for that year would decrease. However, N losses from the plant-soil system would not necessarily follow; and the N not utilized by that crop may be available in following years for crop use. In addition, water-use efficiency could also be affected, if weeds take up a considerable amount of water which would then be unavailable to the crop.

Crop Rotations

The subject of crop rotations for efficient N use is covered in detail by Pierce and Rice in Chapter 3 of this publication. This subject is also dealt with by Kurtz et al. (1984), who pointed out that efficient N use is rarely a major consideration in choosing crop rotations. It is, however, a cornerstone of low-input cropping systems, and does have implications for efficiency in these types of systems.

Nitrogen-fertilizer inputs can be reduced or eliminated by preceding corn (*Zea mays* L.) or a small grain with a leguminous forage crop or green manure. The fertilizer replacement value (FRV) of a legume is the amount of inorganic nitrogen fertilizer required to produce a yield of the subsequent crop equivalent to that produced following the legume (Harris and Hesterman, 1987). Studies investigating the FRV of alfalfa (*Medicago sativa* L.) in rotation with corn have estimated values ranging from 39 to 135 kg of N ha^{-1} (Harris and Hesterman, 1987). The FRV of meadow (composition not known) in a corn-oat (*Avena sativa* L.)-meadow rotation in Iowa ranged from 138 to 224 kg of N ha^{-1} (Sutherland et al., 1961). Fertilizer replacement values can vary greatly depending on the legume being tested, the length of time it is present in rotation, and the management practices used.

Numerous studies have been done on the effect of crop rotation on yields. Many of these studies took place prior to the 1950s, before chemical fertilizers and pesticides came into wide usage.

In long-term experiments at Sanborn field, Missouri, crops in rotation consistently out-yielded those grown continuously (Smith, 1942). This was true even when sufficient N was provided to the crop, indicating that the yield increase was due to other factors in addition to N. Manure treatments were more effective than fertilizer at maintaining soil N levels, and the more often corn was included in a rotation, the more rapid was the decline in total soil N. Long-term soil productivity dropped drastically under continuous corn.

The Morrow plots in Illinois showed similar results. Over a 100-yr period, continuous corn consistently yielded less than corn grown in rotation with oat and red clover (*Trifolium pratense* L.), with or without fertilizer (Odell et al., 1984). Similar results were also obtained in southwestern Ontario (Bolton et al., 1976), Nebraska (Sahs and Lesoing, 1985), Iowa (Voss and Shrader, 1984), and Ohio (Dick and Van Doren, 1985). While some studies indicate that the yield increases in crops grown in rotation are due solely to the N supplied to the crop (Shrader et al., 1966, Baldock and Musgrave, 1980), others demonstrate that yield increases are due to other "rotation effects," such as enhancement of soil physical properties (Odell et al., 1984; Barber, 1972; Voss and Shrader, 1984).

Leguminous Crops and Green Manures

If a forage crop is not a necessary product of a cropping system, legumes can be introduced into a rotation as green manures. Low-input systems often include legumes as green manures or cover crops, in addition to other types of mulches, to provide organic matter and N to the system. Cropping system efficiency may be increased or decreased by these techniques.

In 1964, the Woburn organic-manuring experiment in England (Mattingly, 1973) was begun to investigate the effects of various organic soil amendments on yields and N use of various crops in rotation. Peat, straw, farmyard manure, and trefoil (*Lotus corniculatus* L.) green manure were compared. The green manure greatly increased potato (*Solanum tuberosum* L.) yield, more so than could be accounted for solely by the N provided to the crop. Potatoes were grown during a dry year, and it is suggested that the trefoil residues may have conserved moisture, although no soil moisture measurements were taken. The trefoil may also have provided N to the crop slowly throughout the season. Peat and straw mulches also increased potato yields, and the lowest yields were in those plots that received only chemical fertilizer.

Similar results have been reported by Lockeretz et al. (1980), in which organically managed farms in the midwestern USA experienced lower yield losses under adverse conditions than neighboring conventionally managed farms, and by Sahs and Lesoing (1985), in which corn managed in a rotation that included a legume hay crop suffered less from drought stress than continuous corn. Under low-moisture conditions, treatments involving the addition of organic matter may increase the efficiency of N use by conserving water, increasing the potential for plant growth.

Green manures may also interfere with crop growth, reducing NUE by limiting yield. In the Woburn organic-manuring experiment described above, barley (*Hordeum vulgare* L.) yields were reduced by competition with the trefoil green manure (Mattingly, 1974). In an earlier green manuring experiment at Woburn (Dyke et al., 1977) involving trefoil undersown into barley, the nurse-crop barley that did not receive any additional N fertilizer experienced a yield reduction of 10% as a result of competition from the trefoil. However, the following year the barley yield increase from the green manure (also without additional N fertilizer) outweighed the previous year's loss.

The effect of green manure on crop yield is dependent on management practices, climate, and species of green manure and crop. In Wyoming, the effects of spring-planted alfalfa, Austrian winter pea (*Pisum sativum* spp. *arvense* L. Poir), hairy vetch (*Vicia villosa* Roth) and sainfoin (*Onobychis viciifolia* Scop.), with and without a legume forage harvest, on a succeeding barley crop were compared under conditions of high elevation (> 1280 m above sea level) and a short-growing season (< 148 d) (Abernethy and Bohl, 1987). Uncut hairy vetch, Austrian winter pea, and cut or uncut alfalfa provided barley growth greater than or equal to 100 kg of fertilizer N ha^{-1}. Barley following sainfoin performed better than the barley grown with no fertilizer or legume, but did not perform as well as the barley control treatment that received 100 kg of N ha^{-1}. Alfalfa performed best of all under these conditions, and provided adequate barley growth even with the removal of 3.7 Mg ha^{-1} forage.

Animal Manures

The use of animal manures for N and organic matter is also an integral part of many low-input farming systems. Animal manures can be a cost-effective method of fertilizing a crop, especially if the manure is produced

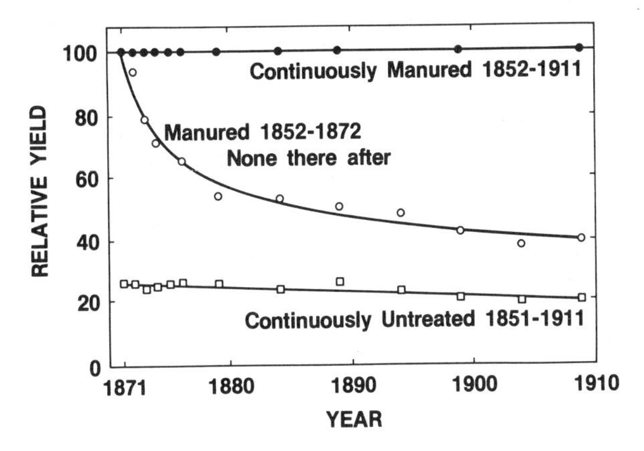

Fig. 12–1. Relative yields of barley over time illustrating the long residual effects of manure (Salter and Schollenberger, 1939).

on-farm. However, N losses from either leaching or volatilization can occur if the manure is applied before the crop can take it up, especially if the manure is applied in the fall or winter (Aldrich, 1984). In addition, Bouldin et al. (1984) estimate that 50% of the N in fresh manure applied in the USA may be volatilized before it is even spread. Animal manure, however, is still a significant source of nutrients.

An important and potentially valuable quality of animal manures is that some fraction of the N contained in them becomes available at a slower rate than fertilizer N. A concept of a "decay series" has been developed to estimate the amount of N released from manure in a given year. An explanation of the decay series concept is contained in Bouldin et al. (1984). The numbers in a decay series vary to represent the variable amount of inorganic nitrogen in different manures available to the first crop, and the gradual mineralization of organic nitrogen in succeeding years.

Long-term positive effects of animal manure have been demonstrated in an experiment at Rothamsted on continuous barley (Salter and Schollenberger, 1939). A plot of continuous barley that had received 31.4 Mg ha^{-1} yr^{-1} manure (fresh wt.) for 20 yr was split in 1872; one-half continued to receive manure yearly, the other received no further manure. Both plots were cropped with barley for another 40 yr. The yields of both treatments are compared with a third plot of continuous barley that had not received any manure from the beginning of the experiment (Fig. 12–1). After 40 yr, the plot that had ceased to receive manure in 1872 was still yielding double compared to the plot that had never received manure. The long-term effects of manure are probably the result of its contribution to the stable organic matter of the soil.

SOIL PROCESSES IN CROPPING SYSTEMS UTILIZING THESE STRATEGIES

Nitrogen loss from the plant-soil system is an important aspect of NUE. This section considers the effects of the management strategies described above on the form and distribution of N in the soil, and N losses from the system, primarily through leaching. The effects on soil structure and biological activity, and the implications of these effects on N losses and NUE, will also be considered.

Nitrogen Availability and Leaching Losses

There have been several studies in recent years utilizing ^{15}N technology to investigate the movement and transformations of N derived from various sources. A study reported by Ladd and Amato (1986) examined the availability and distribution of N derived from medic (*Medicago littoralis* L.), urea, $(NH_4)_2SO_4$, and KNO_3, applied at various rates to soils cropped with two successive wheat crops. Nitrogen sources were labelled with ^{15}N. The experiment took place under field conditions in southern Australia, a semiarid area, in open-ended steel cylinders installed to a 90-cm depth.

It was found that only 17% of the legume N was taken up the 1st yr by the crop, compared to 46% of the fertilizer N. At the time of sowing the second crop, 62% of the N applied as legume was present as soil organic nitrogen, compared to 29% in the fertilizer-N treatments. Total recovery of ^{15}N in crop and soil at the time of sowing the second wheat crop (total ^{15}N in soil to 90-cm depth added to the amount removed in tops of the first crop) was 80 and 84% in the fertilizer and legume plots, respectively. In terms of the 1st-yr wheat crop, the legume treatment was less efficient than the fertilizer treatments in providing N to the crop. However, by the time of sowing the second crop, slightly more N had been lost from the soil-crop system in the fertilizer treatments. Over a longer period of time, legume sources of N may prove to be more efficient than fertilizer sources, since leaching losses may be reduced because of the slowly available nature of the N derived from these sources. More research needs to be done in this area, to determine whether inefficiencies in 1st-yr utilization of N may be translated into longer-term efficiencies.

A similar experiment in Michigan investigated the recovery of ^{15}N from alfalfa residue by a subsequent corn crop (Harris and Hesterman, 1987). Treatments were applied to plots at a rate equivalent to 112 kg of N ha^{-1}. Whole corn plants recovered 20.8% of alfalfa-^{15}N, averaged across both sites in the experiments. Assuming a typical quantity of alfalfa dry matter of 6048 kg ha^{-1} incorporated in an alfalfa-corn rotation, a contribution of 18 to 27 kg of N ha^{-1} by the alfalfa was estimated. This figure is well below the FRV estimates presented earlier in this chapter, suggesting that FRV figures may overestimate the N contribution of alfalfa. The authors conclude that "the main value of the legume appears to be long-term, that is, to maintain levels of soil N high enough to ensure adequate delivery to future crops". This concurs with Ladd (1981), who made a similar conclusion based on an experiment on the availability of medic-N to wheat.

These and other studies using ^{15}N should be interpreted carefully, since a certain amount of underestimation of N uptake is inherent in this technique. Nitrogen transformation in soil has been thoroughly reviewed by Stevenson (1986), who states that (p. 183) ". . .plant uptake of applied ^{15}N does not necessarily provide a true measure of fertilizer N efficiency. Because of turnover by mineralization-immobilization, some of the soil N not otherwise available is taken up by the plant and a corresponding amount of fertilizer N is immobilized." This explains why the conventional method of determining fertilizer-use efficiency from the difference in crop uptake between a fertilized plot and an untreated plot gives higher recoveries of fertilizer N than does the tracer method. It is not clear at this point if this underestimation of N uptake will be similar between legume-N sources and fertilizer N sources, since soil biological processes are responsible for the error, and the organisms responsible will likely respond differently to these radically different N sources.

In other experiments investigating the availability of legume vs. fertilizer N, differences in the timing of N availability have been found. In a pot experiment utilizing ^{15}N that compared the availability of N derived from *Crotalaria* spp., a tropical legume, and fertilizer N to rice (*Oryza sativa* L.), it was found that while the rice absorbed more fertilizer N than legume N initially, after tillering the two sources were absorbed at approximately the same rate, indicating equal availability (Huang et al., 1981). When the two sources of N were combined in a treatment, N recovery (in soil and plant) was found to be intermediate between the treatments of each amendment alone. Nitrogen loss from the rice-soil system was estimated to be about 15, 20, and 27% in the legume, the combination, and the fertilizer treatments, respectively.

These results have important implications for NUE. While legumes may be less-efficient N sources in Year 1 as compared to fertilizer, there may be less N loss from the plant-soil system where legumes are used. This may then translate to greater efficiency in the long term.

Nitrate leaching in the late fall as a result of applications of animal manures or leguminous crop residues has been used as an argument against the use of these amendments (Aldrich, 1984). The techniques used to manage organic amendments or leguminous crops in rotation are crucial for reducing potential leaching losses. Climate and net drainage in the soil are also important factors to take into consideration.

In England, fall plowing of summer hay crops in preparation for winter wheat (*Triticum aestivum* L.) is common practice. High rainfall in the fall and winter months increases net drainage from the soil, increasing the leaching potential of nitrates released by cultivation. An experiment at the Elm Farm Research Centre (EFRC) in England measured leaching on two organic farms as well as in research plots at EFRC (Stopes, 1987). A hay crop of either alfalfa (farm no. 1) or a grass-clover mix (farm no. 2 and EFRC plots) was tilled in September, and winter cereals were planted in October. At the end of December, NO_3-N concentrations in the top 90 cm of soil were 300, 210, and 120 kg ha^{-1} on farm no. 1, farm no. 2, and EFRC, respectively. Leaching loss after alfalfa was estimated to be up to 220 kg of N ha^{-1} by the end of March. On the EFRC plots (grass-clover mix) leaching losses may have been up to 100 kg of N ha^{-1} by the following spring.

An experiment in New Zealand (660 mm yr^{-1} mean annual rainfall) investigated leaching losses in a four-phase rotation of white clover (*Trifolium repens* L.)-pea (*Pisum* spp.)-wheat-wheat (Adams and Pattinson, 1985). Leaching losses from the top 100 cm of soil were lowest under the white clover hay, at 10 kg ha^{-1}. Nitrification increased when the hay was plowed under in late May (late fall). The highest-leaching loss, 90 kg of N ha^{-1}, was during the pea phase of the rotation. This N was released by the previous year's clover crop as well as by the pea residues. Leaching losses under the wheat crops were calculated at 60 and 35 kg of N ha^{-1} for the 1st and 2nd yr wheat crops, respectively.

A 5-yr agroecosystem study at Uppsala, Sweden (Long and Hall, 1987) measured C and N flows in cropping systems that included alfalfa hay, fescue (*Festuca pratensis* Huds.) hay receiving 200 kg of N ha^{-1} yr^{-1}, and fertilized (120 kg of N ha^{-1} yr^{-1}) and unfertilized spring barley (*Hordeum distichum* L.). The hay and barley plots were maintained for 4 and 6 yr, respectively. It was found that alfalfa fixed about 384 kg of N ha^{-1} yr^{-1}, and about 95% of this was retained in the soil-plant system. In addition to flow between soil animals, microbes, the soil organic matter pool, and soil mineral N pool, N losses from the system were also measured. Leaching losses were greatest from the fertilized cereal system at 18 kg of N ha^{-1} yr^{-1}, amounting to 15% of the N added as inorganic fertilizer. Reduction in fertilizer input did not completely remove this problem however, as the barley without any added fertilizer still lost 10 kg of N ha^{-1} yr^{-1}. Losses from both the N_2-fixing hay and the fertilized grass crop were an order of magnitude lower (1 kg of N ha^{-1} yr^{-1}), suggesting that hay crop ecosystems are much more effective at retaining N, even though amounts of N in all components of these ecosystems are higher than in the cereal systems. They caution, however, that leaching losses into drainage water are increased more by the conversion from hay to cereal than by fertilization, at least in the short term.

This conclusion is also supported by a study in Wisconsin (Olsen et al., 1970) which demonstrated that total NO_3 concentration in the soil profile was positively correlated to the frequency of corn in the rotation and higher rates of N application. Again, most leaching occurred between fall and spring, rather than during the growing season.

The results from both Sweden and New Zealand suggest that leaching is only a threat when the legume crop is plowed down prior to a row crop. Alternative systems need to be developed to minimize these losses. Cultivating the hay in the spring for a spring grain or planting a winter cover or "catch" crop to take up mineralized N in the fall are two alternatives being investigated at Elm Farm Research Centre (Stopes, 1987). Aldrich (1984) states that winter grains such as wheat, barley, rye (*Secale cereale* L.), and ryegrass (*Lolium* spp.) tie up only 22 to 34 kg of N ha^{-1}, and thus are ineffective as cover crops. In the New Zealand experiment (Adams and Pattinson, 1985), wheat was estimated to take up about 60 kg of N ha^{-1} in June through August, although this amount still was not enough to eliminate the leaching of high concentrations of N resulting from the two previous legume crops. Data from the National Seed Development Organization, England (Stopes, 1987), shows a range of 34 to 116 kg of N ha^{-1} taken up by wheat planted between early September and late October, respectively, sampled in late March. Early planting of wheat can potentially limit leaching by increasing plant uptake. It is also important to choose appropriate species for use as cover crops.

A lysimeter study in Kentucky (Karraker et al., 1950) investigated leaching losses under various cover crops. Winter leaching losses were low under bluegrass (*Poa* spp.), rye, and alfalfa, as well as under healthy stand of white and red clover. When the clover winter-killed, however, leaching losses were high. Leaching was high under lespedeza (*Lespedeza* spp.), but most of the leaching was prevented when a rye cover crop was grown with the lespedeza.

Research has shown both soybean (*Glycine max* L. Merr) (Johnson et al., 1975) and alfalfa (Schertz and Miller, 1972) to be efficient scavengers of mineralized N in the soil profile. Alfalfa may be especially useful in this respect, because its deep rooting system can scavenge N deeper than many other crops (Schertz and Miller, 1972). More research into the use of various cover crops for N conservation in winter months is needed, as well as other management practices for conservation of N, such as spring plowing and early planting.

Soil Structure and Microbial Activity

Soil structure and microbial activity have important implications in N cycling and water use, and ultimately in both water and N-use efficiency. It is well established that there is a positive correlation between soil organic matter and soil microbiological activity (Martyniuk and Wagner, 1978; Power and Doran, 1984; Schnurer et al., 1985). Soil organic matter tends to be greater in cropping systems utilizing crop rotations than in continuous cropping, and in systems utilizing animal manures rather than commercial fertilizers (Martyniuk and Wagner, 1978; Sahs and Lesoing, 1985). Long-term use of green manures also increases soil organic matter (Reganold et al., 1987). There is some evidence suggesting that increased soil biological activity in cropping systems amended with animal manure may be related to factors other than soil organic matter (Power and Doran, 1984).

Although the connection between soil organic matter and microbiological activity is fairly clear, the connection between these factors and soil structure or resource-use efficiency is less certain. Organic matter has been attributed to encouraging soil granulation, water storage, soil organism activity, nutrient supply and productivity (Reganold et al., 1987). Improved soil structure stemming from increases in organic matter may improve soil aeration (Barber, 1972), which could reduce denitrification losses of N that occur under anaerobic conditions. On the other hand, in a review of the effects of green manuring on physical properties of soils, MacRae and Mehuys (1985) concluded that (i) organic matter does not directly affect all physical parameters, (ii) the effects of green manuring depend on plant species, (iii) green manures do not necessarily improve soil physical condition, and (iv) crop performance may not improve with better soil physical condition.

A comparison was made of the long-term effects of organic and conventional cropping on soil structure and soil erosion in the Palouse region of Washington state (Reganold et al., 1987). The "organic" farm was cropped in a rotation of winter wheat, spring pea (*Pisum sativum* L.), and Austrian winter pea included as a green manure, with no fertilizer and minimal pesticide use. The "conventional" farm was cropped in a winter wheat-spring pea rotation with recommended rates of fertilizer and pesticide. Both farms were first cultivated in 1908 or 1909, but the conventional one did not receive fertilizers and pesticides until mid-century.

The organically farmed soil had higher organic matter and polysaccharide contents, and a lower modulus of rupture (an index relating to the hardness of surface crusting). The moisture content of the organically farmed soil was also higher, due to the increased organic matter. Bulk densities were not significantly different between the soils. Topsoil in the organically farmed soil was 16-cm deeper than in the conventionally farmed soil, as a result of greater water erosion in the latter farm. Soil loss due to rill erosion was estimated to be 8.3 Mg ha^{-1} yr^{-1} on the organic farm and 32.4 Mg ha^{-1} yr^{-1} on the conventional farm. In an earlier study on these same farms, microbial biomass and levels of several enzymes were higher in the surface soil of the organic farm as compared to the conventional farm (Bolton et al., 1985).

Whole-farm studies are valuable for pointing out major differences in nutrient management, N cycling, and rotation effects on soil properties, but replicated field plots are necessary to be sure that site differences do not confound the results, and also to obtain better precision when measuring differences in systems.

RODALE RESEARCH CENTER LOW-INPUT CROPPING EXPERIMENT

Beginning in 1981, the Conversion/Farming Systems Experiment was conducted by the Rodale Research Center (RRC) on 6 ha of leased land in Berks County, Pennsylvania. Prior to 1981, wheat and corn were grown on this land using standard fertilizers and pesticides. The soil was primarily Comly silt loam (fine-loamy, mixed, mesic Typic Fragiudalf) with smaller areas of Berks shaly silt loam (loamy skeletal, mixed, mesic Typic Dystrochrept) and Duffield silt loam (fine-loamy, mixed, mesic, Ultic Hapludalf). These soils had high levels of P and K, a pH of 6.7 and 24 g kg^{-1} of organic matter.

The primary design of the experiment consisted of three cropping systems with three starting points (crops) into each of the 5-yr rotations (Table 12–1). The experimental design is described further in Radke et al. (1987). The objectives of the first 5 yr, the conversion part of the experiment, were: (i) to

Table 12-1. Rotation sequences for the 5-yr Conversion Experiment. Each of the three systems has three entry points into the rotations.

Treatment	Year				
	1981	1982	1983	1984	1985
			Low-input with animals, System 1		
1-1	Spring oat Red clover	Red clover	(Manure) Corn	Soybean	(Manure) Corn silage
1-2	Corn	Soybean	(Manure) Corn silage	Wheat Red clover†	Red clover
1-3	(Manure) Corn silage	Wheat Red clover†	Red clover	(Manure) Corn	Soybean
			Low-input cash-grain, System 2		
2-1	Spring oat Red clover	Corn	Spring oat Red clover	Corn	Soybean
2-2	Soybean	Spring oat Red clover	Corn	Wheat Hairy vetch	Corn
2-3	Corn	Soybean	Spring oat Red clover	Corn	Spring oat Red clover
			Conventional cash grain, System 3—Control		
3-1	Corn	Corn	Soybean	Corn	Soybean
3-2	Soybean	Corn	Corn	Soybean	Corn
3-3	Corn	Soybean	Corn	Corn	Soybean

† Overseeding.

define yield-limiting factors that occur during the transition to low-input cropping systems; (ii) to identify methods of minimizing yield reductions; and (iii) to identify some of the basic processes that take place during the conversion. Two of the cropping systems used no commercial fertilizers or pesticides and are referred to as *low-input* (LIP). System 1 was called *low-input with animals* (LIP-A) since animal manures were applied prior to planting the corn crops in the rotation. System 2, *low-input cash grain* (LIP-CG), did not receive animal manures and relied on plowdowns of green-manure legume crops for nutrients and organic matter. System 3 was a conventional cash grain (CONV) system growing a rotation of corn and soybean, using Pennsylvania State Univ. recommendations for fertilizers and pesticides.

The Farming Systems Experiment began in 1986 as a continuation of the Conversion Experiment (Table 12-2). Multiple-cropping techniques based on earlier work described by Janke et al. (1987) were used in the LIP-CG system to get five (four saleable) crops in 3 yr. This was accomplished by relay cropping soybean into a small grain. In both the LIP-CG and the LIP-A systems, green manures were established by broadcast seeding red clover into a standing small grain. Using this multiple cropping technique, a green manure crop was produced without sacrificing a year of cash crop production.

Nitrogen

Corn and soybean were the two crops grown in all three of the cropping systems. This section will concentrate on the effects of N on the corn crops

Table 12-2. Rotation sequences for the 5-yr Farming Systems Experiment. Each of the three systems has three entry points into the rotations. RC-Alfalfa is a red clover, alfalfa mixture and corn (SS) is short-season corn.

Treat- ment	Year 1986	1987	1988	1989	1990
			Low-input with animals, System 1†		
1-1	Wheat RC-Alfalfa‡	RC-Alfalfa	(Manure) Corn	Soybean	(Manure) Corn silage
1-2	(Manure) Corn	Soybean	(Manure) Corn silage Winter wheat	Winter wheat RC-Alfalfa‡	RC-Alfalfa
1-3	(Manure) Corn silage Winter wheat	Winter wheat RC-Alfalfa‡	RC-Alfalfa	(Manure) Corn	Soybean
			Low-input cash-grain, System 2‡		
2-1	Oat RC-Alfalfa	Corn	Spring barley Soybean Winter wheat‡	Winter wheat RC-Alfalfa‡	Corn (SS) Winter wheat
2-2	Spring barley Soybean Winter wheat‡	Winter wheat RC-Alfalfa‡	Corn	Spring barley Soybean Winter wheat‡	Winter wheat RC-Alfalfa‡
2-3	Corn (SS) Winter wheat Winter wheat	Winter wheat Soybean Winter wheat‡	Winter wheat RC-Alfalfa‡	Corn	Spring barley Soybean
			Conventional cash-grain, System 3—Control§		
3-1	Corn	Corn	Soybean	Corn	Soybean
3-2	Soybean	Corn	Corn	Soybean	Corn
3-3	Corn	Soybean	Corn	Corn	Soybean

† Corn and soybean planted in 76-cm rows. Weed control—Rotary hoe and cultivation for corn and soybean.
‡ Overseeding.
§ Corn planted in 76-cm rows. Soybean drilled into small grain in 18-cm rows. Weed control—Rotary hoe and cultivation in corn.
¶ Corn planted in 76-cm rows. Soybean drilled in 18-cm rows.

since corn is the crop most demanding of N and the crop that suffers the most if N is not available in sufficient amounts at the proper time.

Soil Nitrogen

Soil nitrates in the tilled layer were determined several times throughout the 1985, 1986, and 1987 growing seasons. Soil nitrates in the 0- to 15- and 15- to 30-cm depths for each of the three farming systems in 1985 are shown in Fig. 12-2. A manure application of 176 kg of N ha^{-1} in the LIP-A system provided relative high NO_3 concentrations until Day 163 (12 June) but concentrations declined considerably by Day 200 (19 July). Nitrate concentrations in the LIP-CG system, that had received 180 kg of N ha^{-1} in the aboveground material of hairy-vetch plowed under Day 128 (8 May), increased to high levels on Day 163 and were still high on Day 200. The CONV system received 11 kg ha^{-1} of starter fertilizer at planting (Day 119, 29 April)

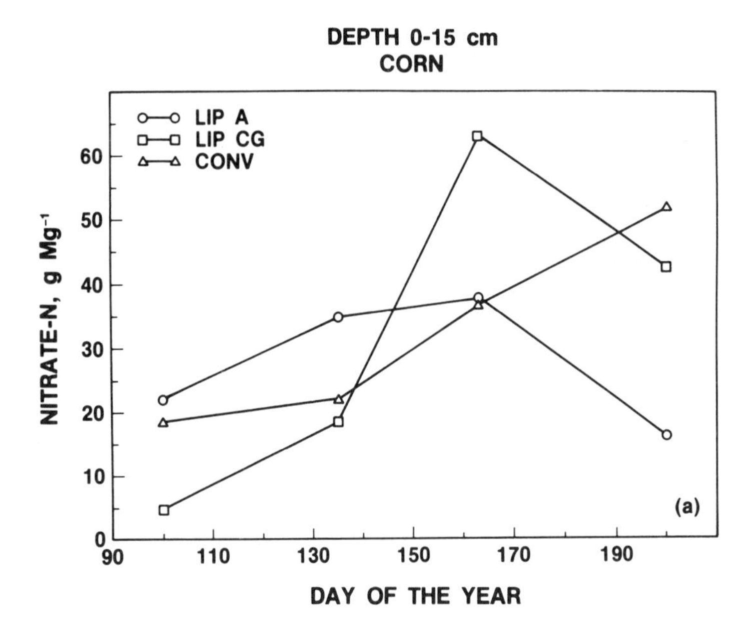

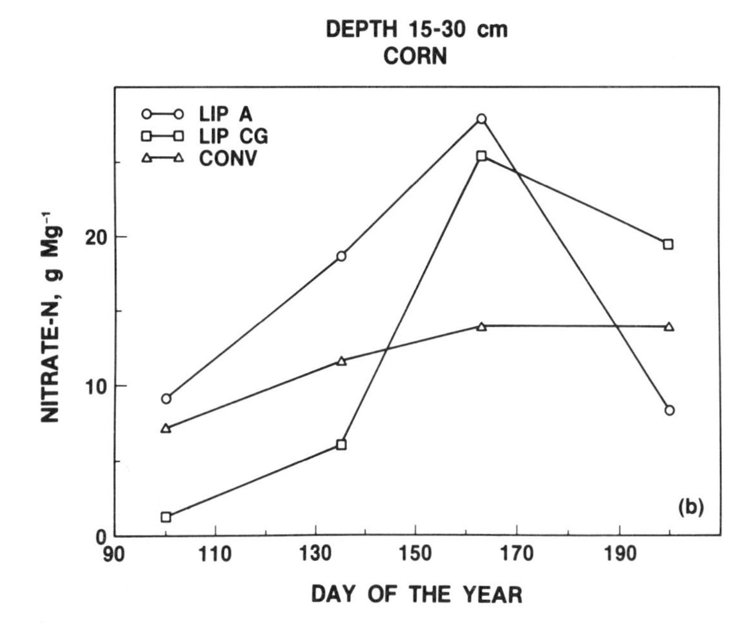

Fig. 12–2. Soil NO_3 concentrations in the corn treatments of the three cropping systems in 1985 for soil layers of (a) 0- to 15-cm and (b) 15- to 30-cm depth. (John Doran, 1987, personal communication. See also Doran et al., 1987.)

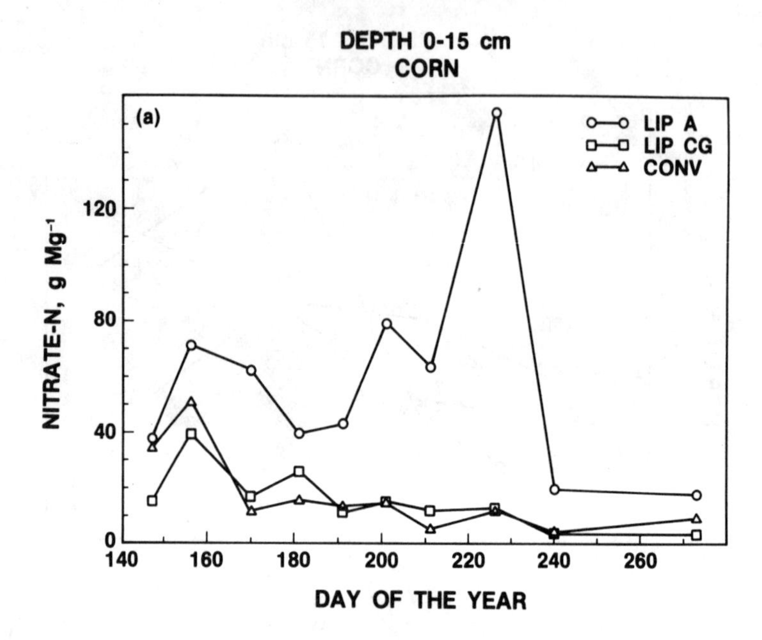

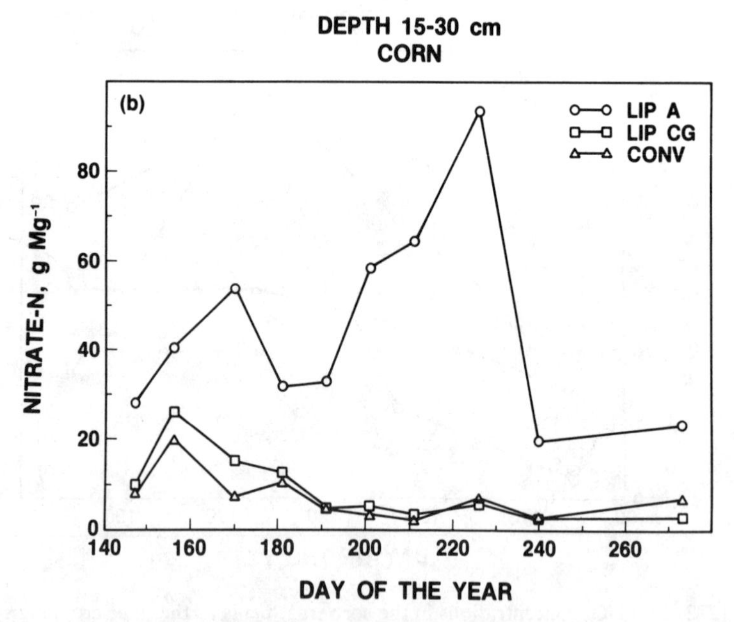

Fig. 12-3. Soil NO_3 concentrations in the corn treatments of the three cropping systems in 1986 for soil layers of (a) 0- to 15-cm and (b) 15- to 30-cm depth.

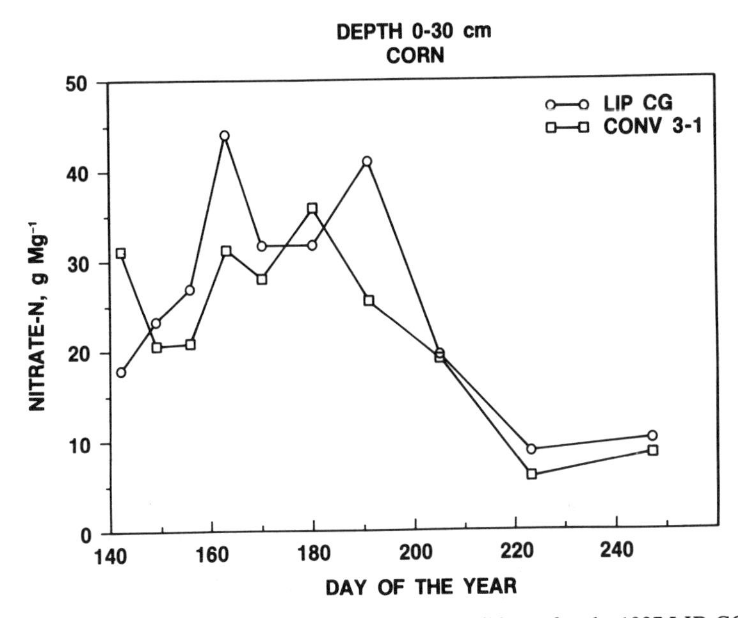

Fig. 12–4. Soil NO_3 concentrations in the 0- to 30-cm soil layer for the 1987 LIP-CG and CONV corn treatments.

followed by a sidedressing of 134 kg ha^{-1} on Day 168 (17 June). This is apparent in Fig. 12–2 by the high NO_3 concentrations on Day 200. Other soil NO_3 determinations in the tilled layers on Day 225 (13 August) showed low concentrations for all three systems (data not shown).

In 1986, relatively high NO_3 levels were sustained in the LIP-A system throughout the first half of the growing season (Fig. 12–3). This was followed by a dramatic increase on Day 226 (14 August) and a subsequent rapid decline. This curve represents the average of two LIP-A treatments that received 421 and 558 kg N ha^{-1} in the form of chicken (*Gallus gallus domesticus*) manure. This was obviously an over-application of N that did not result in higher yields. Soil NO_3 levels in the LIP-CG and CONV corn treatments were similar throughout the season.

Soil NH_4 concentrations were also determined for the corn treatments in each of the three systems. In 1986, there were few differences except at the beginning of the season when the chicken manure applications released more NH_4 earlier in the season (data not shown). NH_4-N concentrations in all three systems peaked between Days 156 and 170 at 8 to 10 g Mg^{-1} in the 0- to 15-cm layer and 7 to 9 g Mg^{-1} in the 15 to 30 cm layer.

Soil NO_3 concentrations were determined in the 0- to 30-cm layer in the LIP-CG and CONV corn treatments in 1987 (Fig. 12–4). The soil NO_3 concentrations for the CONV corn reflects the applications of starter and sidedressed fertilizer on Days 121 (1 May) and 167 (16 June), respectively. The reasons for the two peaks on the LIP-CG NO_3 curve are probably related to weather patterns. There was significantly more NO_3 in the soil profile in the LIP-CG as compared to the CONV system on Days 163 (12 June) and 191 (10 July), and similar NO_3 levels during the rest of the season.

Crop Yield and Leaf Tissue Nitrogen

Corn yields increased in the two LIP systems from 1981 through 1985 (Table 12–3). In 1981, corn yields were 63 and 56% of the CONV corn in the LIP-A and LIP-CG systems, respectively. However, corn yields were low in all three systems in 1981 due to late planting as well as to dry weather conditions. In 1986, corn yield in the LIP-CG system was 82% of CONV, and LIP-A and CONV corn yields were not significantly different. Lower corn yields in the LIP-CG system in 1982, 1983, 1984, and 1986 were due to shorter-season varieties of corn being planted in that system. In 1987, the LIP-CG corn yield was not significantly different from CONV.

Corn leaf tissue N (LTN) concentration in ear leaves collected at silking also increased in the LIP systems, although this was not a consistent trend (Table 12–3). In the LIP-CG system, corn LTN concentration was similar in 1982 through 1984. The high value in treatment 2–2 in 1985 was a result of an especially vigorous stand of hairy vetch being plowed under. In 1986 and 1987, treatments 2–3 and 2–1 increased from 1984 values of 26 to 28 g kg^{-1}. Both followed a rotation that included at least two green manure crops during the previous 5 or 6 yr.

Corn LTN in the LIP-A system showed a distinct difference when comparing treatment 1–2, which started its rotation with unmanured corn, with 1–3, which started with a manured corn silage crop. In 1981, treatment 1–2 was clearly N deficient, while treatment 1–3 LTN concentration was much higher, although still marginal. Treatment 1–3 corn LTN concentration increased from 1981 to 1984 and again in 1986. In 1983, treatment 1–2 corn, which received more than 200 kg of N ha^{-1}, was even more N deficient than in 1981, and dry matter yield was only 60% of CONV (data not shown). This corn was preceded by unmanured corn in 1981 and soybean in 1982. The other 1983 LIP-A corn (treatment 1–1) received about the same amount of manure as treatment 1–2 but was preceded by small grain and hay resulting in an LTN concentration of 27.3 g kg^{-1}, well above that of all other corn treatments in that year, including CONV. In 1986, all LIP corn treatments had a significantly higher LTN concentration than the CONV corn treatments. There were no significant differences in 1987.

In this experiment, utilization of N, as indicated by corn LTN concentration, was reduced in the LIP-A system if insufficient N was available to corn when the rotation was initiated. Leaf tissue N concentration was improved when corn was not planted until the 3rd yr of the rotation.

Nitrogen-yield Efficiency

Nitrogen-yield efficiency (NYE) was calculated by dividing either corn grain yield or dry matter yield by the amount of N contained in legume tops,

Table 12–3. Corn grain yields, leaf tissue N, and N applied as animal manure (LIP-A), green manure (LIP-CG), and commercial fertilizer (CONV) for the corn treatments in the Conversion/Farming Systems Experiment.

System	Treatment	Grain yield	Leaf tissue N	Nitrogen applied
		kg ha^{-1}	g kg^{-1}	
		1981		
LIP-A	1–2	1 559b*	19.8c	0
LIP-A	1–3	Silage	24.1b	179
LIP-CG	2–3	1 374b	20.3c	0
CONV	3–1	2 374a	28.7a	179
CONV	3–3	2 539a	29.5a	179
		1982		
LIP-CG	2–1†	5 467b	26.0a	136
CONV	3–1	9 551a	27.4a	140
CONV	3–2	8 586a	26.6a	140
		1983		
LIP-A	1–1	5 518ab	27.3a	223
LIP-A	1–2	Silage	19.0d	211
LIP-CG	2–2†	4 602b	24.9b	115
CONV	3–2	5 917a	22.0c	134
CONV	3–3	6 347a	23.9b	134
		1984		
LIP-A	1–3	8 697b	25.8b	193
LIP-CG	2–1	7 410c	26.1b	110
LIP-CG	2–3†	5 958d	25.8b	131
CONV	3–2	9 690a	29.8a	145
CONV	3–3	9 330ab	29.8a	145
		1985		
LIP-A	1–1	Silage	25.0c	176
LIP-CG	2–2	9 573a	31.7a	180
CONV	3–2	8 538a	29.2b	145
		1986		
LIP-A	1–2	10 800a	27.9a	421
LIP-A	1–3	Silage	28.0a	558
LIP-CG	2–3†	8 600b	27.8a	48
CONV	3–1	10 550a	26.2b	146
CONV	3–3	10 420a	25.2b	146
		1987		
LIP-CG	2–1	9 032a	28.3a	43
CONV	3–1	8 184a	28.0a	130
CONV	3–2	8 709a	28.7a	130

* Values for each parameter and for each year followed by the same letter are not significantly different at $P = 0.05$.

† Indicates a shorter-season variety was used in the LIP-CG treatment than in the other treatments in the same year.

Table 12-4. Nitrogen-yield efficiency by (NYE) corn (A) grain yield and (B) dry matter.†

Cropping system	Previous crop‡	1981	1982	1983	1984	1985	1986	1987
		kg ha⁻¹/kg ha⁻¹						
		A. NYE by grain yield (grain yield/N supplied to crop)						
LIP-A	Hay	0§	--	24.7	45.1	--	25.7	--
	Soybean	--	--	--	--	--	--	--
LIP-CG	Legume	0	40.2	40.0	67.4¶	53.2	179.2	209.9
	Legume				45.5#			
CONV	Corn	13.3	68.2	44.2	64.3	--	72.3	61.0
	Soybean	14.2	61.3	47.4	66.8	58.9	71.4	64.9
		B. NYE by dry matter (dry matter/N supplied to crop)						
LIP-A	Hay	0	--	52.7	83.0	--	NA††	--
	Soybean	52.6	--	37.7	--	79.3	NA	--
LIP-CG	Legume	0	71.3	80.3	132.3¶‡‡ 109.7#‡‡	99.9	NA	367.8
CONV	Corn	75.6	99.6	97.5	121.2	--	NA	104.0§§
	Soybean	78.7	77.9	101.9	121.4	111.4	NA	127.7

† Nitrogen supplied to crop: N content of animal manure incorporated in the LIP-A system, legume plowed under just prior to corn planting in the LIP-CG system, and fertilizer N applied to the CONV treatment.
‡ Crop preceding 1981 was wheat in all treatments.
§ Zero values in 1981 in the LIP-A and LIP-CG systems, respectively, indicate no animal or green manure was plowed under prior to planting.
¶ Treatment 2-1, full-season corn.
# Treatment 2-3, short-season corn.
†† Data not available.
‡‡ One replication eliminated from each of the LIP-CG dry matter results due to destruction of sample during processing.
§§ Follows soybean crop.

animal manure, or N fertilizer applied to the crop (Table 12-4). The NYE figures given for the LIP-CG system do not account for N supplied by the green manure roots or winter-killed tops from the previous year. This makes the figures for this system artificially high. In spite of this, there are some interesting trends.

In the LIP-A system, NYE figures were generally low relative to the other two systems. In 1984 and 1985 they were higher than in earlier years, but in 1986 they dropped because of the overapplication of chicken manure. The low figures in 1981 and 1983 were probably the result of lower availability of manure N as compared to fertilizer N. The manure put down in 1981 had a lower N content than manures used in subsequent years. In addition, all cropping systems in 1981 had low NYE figures, especially for grain yield, due to low yields caused by dry weather and late planting. In 1984 and 1985, there was a greater response to the manure, probably indicating some mineralization of the N contributed from manure in previous years.

Nitrogen-yield efficiency in the LIP-CG system tripled in 1986, and increased again in 1987. The total N plowed under as green manure in these years was small. Not only was the percentage N of the green manure low due to a large proportion of nonleguminous weed residues, but the total

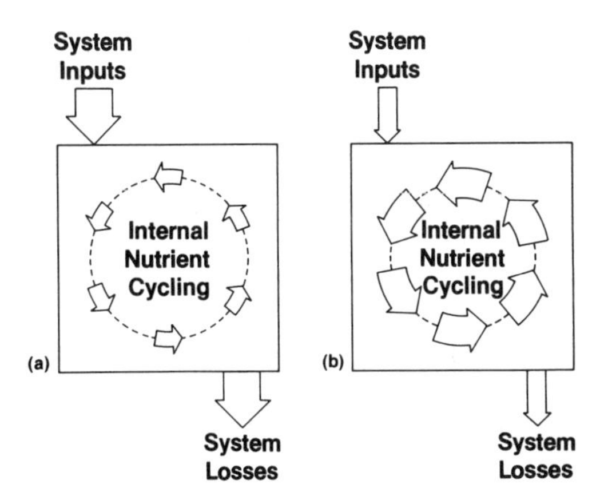

Fig. 12–5. A conceptual model of (a) a cropping system requiring relatively large N inputs to sustain crop yield as compared to (b) one characterized by intensified N cycling to sustain crop yield.

biomass was lower because the crop was plowed under earlier in the season than in previous years, to accommodate a full-season variety of corn. The sharp increase in NYE probably reflects green manure-N that was plowed down in previous years becoming slowly available. The ability of the LIP-CG system to sustain these levels of efficiency is questionable, but is seems that with continued inputs of greater quantities of green manure, commensurate with the first 5 yr, higher NYE levels than for the CONV corn could be maintained.

Discussion

Nitrogen availability to the corn crop was perhaps the most crucial factor in the RRC low-input cropping systems. The reduced corn yields and generally lower LTN concentrations in the LIP systems as compared to CONV indicate the need for improved soil N status during the early part of the rotation. In contrast, LIP soybean yields and LTN concentration were always equal to or slightly higher than for CONV soybean (data not shown). Small grains and forage legumes occurred only in the LIP systems but LTN levels were sufficient and yields were comparable to Berks County averages (data not shown).

Operations such as manuring, tillage, and planting, and their effects on the timing of N availability, are important for maintaining adequate yields while minimizing potential pollution problems. In order to optimize NUE in low-input systems, a greater understanding of the biological, chemical, and physical processes occurring in low-input systems is needed.

Empirical evidence from the Rodale Conversion/Farming Systems Experiment and a growing body of literature indicate that biological processes are responsible for striking differences between conventional and low-input systems in terms of internal-N cycling. It appears that the ''active'' fraction of soil organic matter may be responsible for these differences, as illustrated in Fig. 12–6. In conventional systems, relatively large inputs of mineral N uired each season to produce a corn crop, as shown in Fig. 12–5a. ystem represented by Fig. 12–5b illustrates our hypothesis (greatly plified) that increased N cycling within the system occurs as a result of enhancement of the active soil organic matter fraction following applications of legume and animal manure. In the Rodale Conversion/Farming Systems Experiment, increased N cycling may have been the reason for equal corn yields from conventional and low-input systems in 1986 and 1987, in spite of the relatively low N input from the legumes during these years.

Mechanistic evidence for this hypothesis is found in data from the 5-yr experiment conducted in Uppsala, Sweden, already reviewed in this chapter, comparing C and N flow in cereal vs. hay-cropping systems. Long and Hall (1987) report that, "The cereal systems show considerably lower fluxes of N between soil components, reflecting a lower input of this element into the soil as organic matter, and reduced microbial consumption and mineralization in the cereal system."

This same concept has also been suggested by Patriquin (1986), as a result of measurements of N cycling on a diversified farm in Nova Scotia, Canada. Nitrogen deficiency in cereal crops was observed during the transition from fertilizer to green manure sources of N, but it was found that, "the problem lay not in the quantities of N entering the system, but in the way it cycled around the system." Patriquin proposes that "productivity is intensified not by augmenting inputs but by intensifying cycling. This is achieved by maximizing the biological activity of all components of the systems."

More evidence in support of this hypothesis is found in the classic work of Stanford and Smith (1972) who quantify potentially mineralizable N from various soil types and cropping regimes. The N-mineralization potential in a sugarbeet (*Beta vulgaris* L.)-barley rotation that included animal manure and fertilizer N was twice as great as the same rotation with only fertilizer N. The same study also showed that twice as much N was mineralized from soils under a crop rotation that included corn and 3 yr of fescue as compared to continuous corn.

Using a similar methodology with the soils from the Uppsala, Sweden experiment, Bonde and Rosswall (1987) found that more N mineralized from the grass and legume hay plots than from either barley plot. In addition, they found seasonal differences in potentially mineralizable N as large as the differences between cropping systems. They conclude that this is evidence for the existence of an active fraction of soil organic matter.

Water

Efficient water use implies that the fraction of the available water that goes to transpiration should be maximized (Loomis, 1983; Tanner and Sinclair, 1983). It follows that water-use efficiency can be increased by minimizing the amount of water lost to crops through soil evaporation, runoff, deep percolation, and use by weeds. Water-use efficiency may be increased by more intensive cropping systems, such as multiple cropping, or by cropping over a greater portion of the year. However, water-use effi-

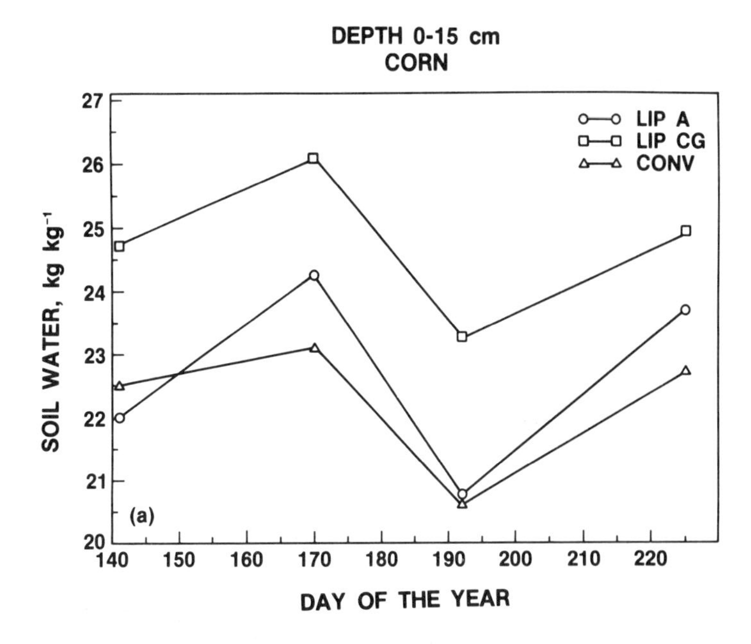

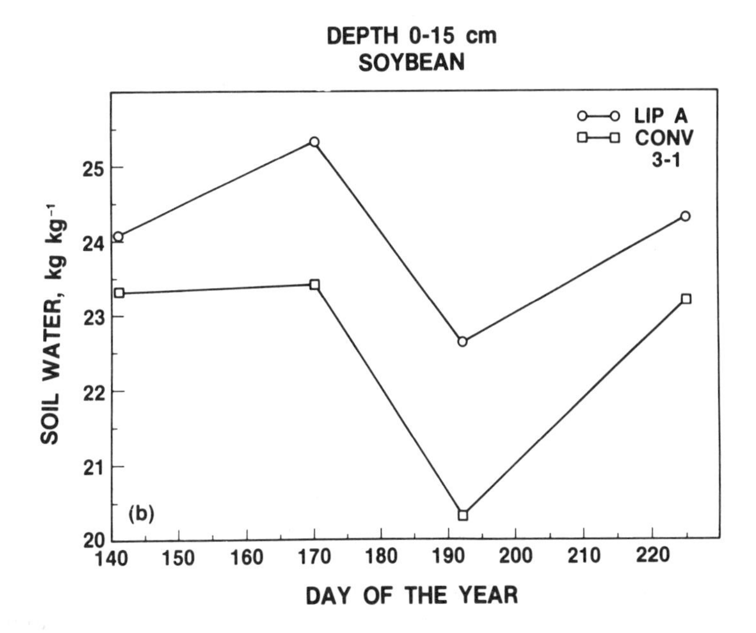

Fig. 12-6. Soil water contents in the 0- to 15-cm layer in 1985 for the (a) corn and (b) soybean treatments.

may be decreased if there is insufficient water for all of the crops ,nis, 1983).

Water in the Soil Profile

In the Rodale Conversion/Farming Systems Experiment, soil water profile contents measured weekly to the 120-cm depth during the growing season were seldom different in the corn or soybean treatments in any of the three cropping systems for the years 1985, 1986, and 1987 (data not shown). Therefore, water use, calculated from precipitation events and weekly soil water profiles, was the same within crops (corn or soybean). However, different crops did show different use patterns. For example, winter wheat used more water in the early growing season when corn and soybean were just starting to grow. Likewise, water-use patterns under multiple cropping in the LIP-CG system were different but total soil-water depletion from May through September did not exceed any of the other treatments. Soil water contents were not measured in the late fall and early spring when winter crops were still using water and nutrients.

Soil Water in the Tilled Layer

Gravimetric soil water contents were determined for the 0- to 15- and 15- to 30-cm depths in 1985 and 1986. Water contents in the corn treatments were significantly higher in the 0- to 15-cm layer of the LIP-CG system than the LIP-A or CONV systems for all four sampling dates in 1985 (Fig. 12–6a). Water contents were not significantly different in the LIP-A and CONV treatments. The results for the 15- to 30-cm layer were similar (data not shown). The LIP-CG corn was grown after a plowdown of live hairy vetch. The hairy vetch not only provided ample N for the 1985 corn crop but, apparently, improved the water-holding characteristics of the tilled layer.

The LIP-A soybean treatment also had more water than the CONV soybean treatments in the 0- to 15-cm soil layer for the four sampling dates in 1985 (Fig. 12–6b). Data were similar in the 15- to 30-cm layer. Both soybean crops followed a corn crop in 1984.

Similar results occurred in 1986 (data not shown). Low-input treatment soils were wetter in the tilled layer than the CONV treatments for most dates. Soil water contents were not different among treatments in the 0- to 15-cm layer during two dry periods around Days 156 and 181 and after Day 240.

Discussion

No definitive statements can be made concerning soil water used in the Rodale cropping systems, except that the tilled layers of the LIP systems tended to be wetter at times. This could be an important factor for microbial activity and its effect on N transformations.

Cropping systems need to be devised to maximize the use of soil-available water without causing periods of water deficit. Multiple cropping may be advantageous in areas with plentiful water, such as eastern Pennsylvania with

> 100 cm of annual precipitation. Cover crops might use soil water and nutrients in the fall and spring that would otherwise be lost through leaching. Cover crops should also increase infiltration and reduce runoff. Further studies are needed on these aspects in LIP cropping systems.

CONCLUSION

The concept of N-use efficiency, as currently defined, needs to be modified to be useful in evaluating LIP systems that rely on legume and animal manure sources of N. Realistic estimations of N availability and loss in these systems will probably involve complicated N budgets that take into account the slow release of organic sources of N, transformations of N in the soil, and cycling processes that involve soil micro- and macro-fauna.

While studies have shown N uptake to be greater from fertilizer sources than legume sources during the year of application, greater efficiency of fertilizer N over a longer period of time has not been demonstrated. In fact, the slower mineralization of N from legume and manure sources may reduce loss from the soil-crop system, increasing efficiency over the course of one or more cycles of a rotation. "Intensified" N cycling through an increase in soil biological activity in low-input systems may also contribute to greater N-use efficiency over the long term. It is important to emphasize that proper management is crucial for maximizing efficiency and minimizing loss and environmental degradation in systems utilizing these nutrient sources. Managed carefully, these crop amendments can be an integral part of an efficient LIP cropping system.

The Rodale cropping systems study is one of a few that have looked at LIP cropping systems designed to make optimal use of internal resources while minimizing external inputs, with a goal of maximizing returns rather than production. The cropping systems were studied holistically, much like farmers would implement them on their farms. Many disciplines were involved in the study and several lessons were learned.

1. Efficiency of N use is increased by starting a LIP rotation with crops that demand less N than does corn. If corn must be planted during the first 3 yr, it should be supplied with N through animal or green manure, and/or small amounts of commercial fertilizer to help maintain yields during the transition. A high C/N ratio in the residues may result in immobilization of a large proportion of N. Proper management is crucial to assure enough N without over-applying, especially in the case of animal manures. Nitrogen-yield efficiency improves over the years as organic soil N stores increase.
2. Multiple cropping is an efficient way to establish green manure crops without sacrificing cash crop production, and looks promising for increasing economic returns from LIP-CG systems by allowing for more intensive crop rotations. It also offers opportunities for enhancing nutrient cycling and weed control, and for reducing erosion and

leaching losses. Multiple cropping may have adverse effects in areas of limited water.

3. To provide good crop yields and high net returns, careful management and planning of LIP cropping systems is essential. A balanced-cropping rotation is needed to assure adequate internal resources, including N, water, and weed control.

ADDITIONAL RESEARCH NEEDS

The Rodale cropping system experiments continue and others are being conducted at several places in the USA and the world. It is important to have studies of LIP systems in many geographical regions, to better understand the effects of different climates, soils, and crops. Specific subjects that need to be addressed within the context of LIP systems include:

1. Studies on the dynamics of decomposition of organic materials and their short- and long-term effects on N cycling.
2. Multiple-cropping techniques which enhance water and N-use efficiency.
3. Rooting characteristics relating to uptake of nutrients and water.
4. The interaction between legume cover crops and tillage practices that optimize N availability, minimize N leaching, and maintain weed, insect, and pathogen control.

Low-input cropping systems have been advocated as a possible solution to many of the environmental and economic problems facing agriculture today. The development and fine-tuning of LIP cropping systems to maximize resource-use efficiency, optimize economic returns, and minimize negative environmental impacts will provide a challenge to researchers of many disciplines.

REFERENCES

Abernethy, R.H., and W.H. Bohl. 1987. Effects of forage legumes on yield and nitrogen uptake by a succeeding barley crop. Appl. Agric. Res. 2:97–102.

Adams, J.A., and J.M. Pattinson. 1985. Nitrate leaching losses under a legume-based crop rotation in Central Canterbury, New Zealand. N.Z. J. Agric. Res. 28:101–107.

Aldrich, S.R. 1984. Nitrogen management to minimize adverse effects on the environment. p. 663–673. *In* R.D. Hauck (ed.) Nitrogen in crop production. ASA, CSSA, and SSSA, Madison, WI.

Baldock, J.O., and R.B. Musgrave. 1980. Manure and mineral fertilizer effects in continuous and rotational crop sequences in central New York. Agron. J. 72:511–518.

Barber, S.A. 1972. Relation of weather to the influence of hay crops on subsequent corn yields on a Chalmers silt loam. Agron. J. 64:8–10.

Bock, B.R. 1984. Efficient use of nitrogen in cropping systems. p. 273–294. *In* R.D. Hauck (ed.) Nitrogen in crop production. ASA, CSSA, and SSSA, Madison, WI.

Bolton, E.F., V.A. Dirks, and J.W. Aylesworth. 1976. Some effects of alfalfa, fertilizer and lime on corn yield in rotations on clay soil during a range of seasonal moisture conditions. Can. J. Soil Sci. 56:21–25.

Bolton, H., Jr., L.F. Elliot, R.I. Papendick, and D.F. Bezdicek. 1985. Soil microbial biomass and selected soil enzyme activities: Effect of fertilization and cropping practices. Soil Biol. Biochem. 17:297-302.

Bonde, T.A., and T. Rosswall. 1987. Seasonal variation of potentially mineralizable nitrogen in four cropping systems. Soil Sci. Soc. Am. J. 51:1508-1514.

Bouldin, D.R., S.W. Klausner, and W.S. Reid. 1984. Use of nitrogen from manure. p. 221-245. *In* R.D. Hauck (ed.) Nitrogen in crop production. ASA, CSSA, and SSSA, Madison, WI.

Buttel, F.H., G.W. Gillespie, Jr., R.R. Janke, B. Caldwell, and M. Sarrantonio. 1986. Reduced-input agricultural systems: Rationale and prospects. Am. J. Alt. Agric. 1:58-64.

Dick, W.A., and D.M. Van Doren, Jr. 1985. Continuous tillage and rotation combinations effects on corn, soybean, and oat yields. Agron. J. 77:459-465.

Doran, J.W., D.G. Fraser, M.N. Culik, and W.C. Liebhardt. 1987. Influence of alternative and conventional agricultural management on soil microbial processes and nitrogen availability. Am. J. Alt. Agric. 2(3):99-106.

Dyke, G.V., H.D. Patterson, and T.W. Barnes. 1977. The Woburn long-term experiment on green manuring, 1936-67; results with barley. p. 119-149. *In* Rothamsted Report for 1976, part 2. Rothamsted Exp. Stn., Harpenden, Hereford, England.

Hallberg, G.R. 1986. From hoes to herbicides: Agriculture and groundwater quality. J. Soil Water Conserv. 41:357.

Harris, G.H., and O.B. Hesterman. 1987. Recovery of nitrogen-15 from labeled alfalfa residue by a subsequent corn crop. p. 58-59. *In* J.F. Powers (ed.) The role of legumes in conservation tillage systems. Soil Conserv. Soc. Am., Ankeny, IA.

Huang, D.M., J. Gao, and P. Zhu. 1981. The transformation and distribution of organic and inorganic fertilizer nitrogen in rice-soil system. T'u Jang Hsueh Pao 18:107-121.

Janke, R.R., R. Hofstetter, B. Volak, and J.K. Radke. 1987. Legume interseeding cropping systems research at the Rodale Research Center. p. 90-91. *In* J.F. Power (ed.) The role of legumes in conservation tillage systems. Soil Conserv. Soc. of America., Ankeny, IA.

Johnson, J.W., L.F. Welch, and L.F. Kurtz. 1975. Environmental implications of N fixation in soybeans. J. Environ. Qual. 4:303-306.

Karraker, P.E., C.E. Bortner, and E.N. Fergus. 1950. Nitrogen balance in lysimeters as affected by growing Kentucky bluegrass and certain legumes separately and together. Kentucky Agric. Exp. Stn. Bull. 557.

Kurtz, L.T., L.V. Boone, T.R. Peck, and R.G. Hoeft. 1984. Crop rotations for efficient nitrogen use. p. 295-306. *In* R.D. Hauck (ed.) Nitrogen in crop production. ASA, CSSA, and SSSA, Madison, WI.

Ladd, J.N. 1981. The use of ^{15}N in following organic matter turnover with specific reference to rotation systems. Plant Soil 58:401-411.

----, and M. Amato. 1986. The fate of nitrogen from legume and fertilizer sources in soils successively cropped with wheat under field conditions. Soil Biol. Biochem. 18:417-425.

Lockeretz, W., G. Shearer, S. Sweeney, G. Kuepper, D. Wanner, and D.H. Kohl. 1980. Maize yields and soil nutrient levels with and without pesticides and standard commercial fertilizers. Agron. J. 72:65-72.

----, and S. Wernick. 1983. Organic farming: Can it help hard-pressed farmers. Independent Banker (Nov.): 15-16, 46-68.

Long, S.P., and D.O. Hall. 1987. Nitrogen cycles in perspective. Nature (London) 329:584-585.

Loomis, R.S. 1983. Crop manipulations for efficient use of water: An overview. p. 345-374. *In* H.M. Taylor et al. (ed.) Limitations to efficient water use in crop production. ASA, CSSA, and SSSA, Madison, WI.

MacRae, R.J., and G.R. Mehuys. 1985. The effect of green manuring on the physical properties of termperate-area soils p. 71-94. *In* B.A. Stewart (ed.) Adv. Soil Sci., Vol. 3. Springer-Verlag New York, New York.

Martyniuk, S., and G.H. Wagner. 1978. Quantitative and qualitative examination of soil microflora associated with different management systems. Soil Sci. 125:343-350.

...y, G.E.G. 1974. The Woburn organic manuring experiment: I. Design, crop yields ...d nutrient balance 1964–72. p. 98–125. *In* Rothamsted Report for 1973, part 2. ...othamsted Exp. Stn., Harpenden, Hereford, England.

...ers, C.F., J. Meek, S. Tuller, and A. Weinberg. 1985. Non-point sources of water pollution. J. Soil Water Conserv. 40:14–18.

Odell, R.F., W.M. Walker, L.V. Boone, and M.G. Oldham. 1984. The Morrow Plots: A century of learning. Univ. of Illinois Agric. Exp. Stn. Bull. 775.

Olsen, R.J., R.F. Hensler, O.J. Attoe, S.A. Witzel, and L.A. Peterson. 1970. Fertilizer nitrogen and crop rotation in relation to movement of nitrate nitrogen through soil profiles. Soil Sci. Soc. Am. Proc. 34:448–452.

Patriquin, D.G. 1986. Biological husbandry and the "nitrogen problem." Biol. Agric. Hortic. 3:167–189.

Pimentel, D., D. Andow, R. Dyson-Hudson, D. Gallahan, S. Jacobson, M. Irish, S. Kroop, A. Moss, I. Schreiner, M. Shepard, T. Thompson, and B. Vinzant. 1980. Environmental and social costs of pesticides: A preliminary assessment. Oikos 34:126–140.

Power, J.F., and J.W. Doran. 1984. Nitrogen use in organic farming. p. 585–598. *In* R.D. Hauck (ed.) Nitrogen in crop production. ASA, CSSA, and SSSA, Madison, WI.

Radke, J.K., W.C. Liebhardt, R.R. Janke, and S.E. Peters. 1987. Legumes in crop rotations as an internal nitrogen source for corn. p. 56–57. *In* J.F. Power (ed.) The role of legumes in conservation tillage systems. Soil Conserv. Soc. Am., Ankeny, IA.

Reganold, J.P., L.F. Elliot, and Y.L. Unger. 1987. Long-term effects of organic and conventional farming on soil erosion. Nature (London) 330:370–372.

Sahs, W., and G. Lesoing. 1985. Crop rotations and manure versus agricultural chemicals in dryland grain production. J. Soil Water Conserv. 40:511–516.

Salter, R.M., and C.J. Schollenberger. 1939. Farm manure. Ohio Agric. Exp. Stn. Bull. 605.

Schertz, D.L., and D.A. Miller. 1972. Nitrate-N accumulation in the soil profile under alfalfa. Agron. J. 64:660–664.

Schnurer, J., M. Clarholm, and T. Rosswall. 1985. Microbial biomass and activity in an agricultural soil with different organic matter contents. Soil Biol. Biochem. 17:611–618.

Shrader, W.D., W.A. Fuller, and F.B. Cady. 1966. Estimation of a common nitrogen response function for corn (*Zea mays*) in different crop rotations. Agron. J. 58:397–401.

Smith, G.E. 1942. Sanborn field: Fifty years of field experiments with crop rotations, manure, and fertilizers. Missouri Agric. Exp. Stn. Bull. 458.

Stopes, C. 1987. Nitrogen leaching in organic ley arable farming systems. New Farmer Grower. 15:22–23.

Stanford, G., and S.J. Smith. 1972. Nitrogen mineralization potentials of soils. Soil Sci. Soc. Am. Proc. 36:465–472.

Stevenson, F.J. 1986. Cycles of the soil. John Wiley and Sons, New York.

Sutherland, W.N., W.D. Shrader, and J.F. Pesek. 1961. Efficiency of legume residue nitrogen and inorganic nitrogen in corn production. Agron. J. 53:339–342.

Tangley, L. 1986. Crop productivity revisited. Bioscience 36:142–147.

Tanner, C.B., and T.R. Sinclair. 1983. Efficient water use in crop production: Research or re-search? p. 1–27. *In* H.M. Taylor (ed.) Limitations to efficient water use in crop production. ASA, CSSA, and SSSA, Madison, WI.

U.S. Department of Agriculture. 1980. Report and recommendations on organic farming. U.S. Gov. Print. Office, Washington, DC.

Voss, R.D., and W.D. Shrader. 1984. Rotation effects and legume sources of nitrogen for corn. p. 61–68. *In* Organic farming: Current technology and its role in a sustainable agriculture. Spec. Publ. 46. ASA, Madison, WI.